Marlon Wesley Machado Cunico, Ph.D.

3D printers and additive manufacturing: the rise of the industry 4.0

Curitiba

Concep3D

2019

Copyright ©2019 by Concep3D

International Cataloging Data in the Publication (ICDP)

C972.d   Cunico, Marlon Wesley Machado,

3D printers and Additive manufacturing: The rise of the Industry 4.0 / Marlon Wesley Machado Cunico; Concep3D Pesquisas Científicas Ltda; Curitiba, 2019,

ISBN: 9781695364981

1.3D Printer; 2. Additive manufacturing 3. Technology.   Título

CDD 620

Systematic Cataloging Index

1. Engineering and applications 620

1st Edition - 2019

Printed in Brazil

©2019 by Concep3D, All RightsReserved

Concep3D Pesquisas Científicas Ltda

298 Pedro Ivo street ap 23

80010-020 Curitiba, Brasil

http://www.concep3d.com

Preface

Over the years, there has been an increased demand for the manufacture of objects and products of greater complexity, leading to the evolution of manufacturing processes. As a result, several technologies have been developed to try to support these market needs.

Among these technologies, we can highlight the 3D printers, which in recent years has been shown a popularization in the global media. As a consequence, many questions and doubts about this technology are still pending, as for example, the problematic about the use of this equipment for armament manufacture.

For this reason, the main goal of this book is to introduce basic concepts about all the main 3D printing technologies, presenting the main technologies found in the market, as well as the main restrictions and applications for each equipments.

Chapter 1 presents a generalized overview of 3D printers concepts, in addition to exposing the main steps and constraints of the process.

It can also be said that this chapter also presents a historical survey on the evolution of these technologies over the years. In this study, we indicate the leading pioneers of additive manufacturing and the patents that made it possible for 3D printers to be what they are today.

On the other hand, the main technologies found in the market are presented in Chapter 2, as well as the concepts behind these technologies, such as materials, manufacturing strategies, strength, advantages and disadvantages.

Among the main technologies presented, we can highlight: liquids based, solids based and powder based. Therefore, the resistance variation, dimensional distortion and manufacturing resolution are compared to indicate the best technology according to the desired application.

In Chapter 3, the concise discussion about the birth of industry 4.0 and how 3D printers collaborate to the rise of this new industry revolution.

Main aspects of previous revolutions and what are the differences with respect to this new revolution are presented. In addition, we also present the key applications of 3D printers for industry 4.0 besides the risks and respective liabilities.

In Chapter 4, the main applications found for these 3D printers are presented, as well as an indication of the most recommended devices for each of these applications.

Benefits that each type of printers provide for each application area are exposed, as well as an indication of the best 3D printer to use for each phase of the product development process.

In order to demystify myths related to 3D printers, Chapter 5 deals with current affairs, controversies, and trends that is found in these technologies. Key topics discussed in this chapter include the controversy over domestic weapons manufacturing, research, innovation, and future trends.

In this chapter, several researches related to additive manufacturing technologies and 3D printers are presented. For example, the development of organ 3D printers and the manufacture of houses and buildings through 3D printers are presented amid one of the most relevant research related to additive manufacturing.

Overall, it is hoped that this book will help to disseminate the fundamentals of 3D printers and clarify subject-related questions for people of all levels of education.

Prof. Marlon Cunico, Ph.D.

Editor

About the Author

Professor Marlon Wesley Machado Cunico, since obtaining a PhD in Mechanical Engineering, has accumulated more than 15 years of experience in statistical process control, product development, machine design, structural and finite element analysis, optimization and numerical modeling, heat transfer and fluid dynamics analysis, machine control, characterization and development of material.

He has been researching vanguard fields, such as additive manufacturing (AM) technologies, for more than 10 years, where he won the international award of outstanding Doctoral Research from Garpa/Emerald. It is possible to highlight his involvement in several industries, such as white-goods, automotive, agriculture, medical manufacturing, mining and high precision machines.

During his career, he has filed eight patents in addition to generating classified innovation in industry. His interdisciplinary approach led to the development of advanced new manufacturing technologies, numerical methods, processes, materials, instrumentation and control systems.

Marlon is CNPQ Researcher in the postgraduate department of Mechanical Engineering at the University of São Paulo, Brazil, where he develop projects related to advanced manufacturing for medical applications, such as medical equipment, advanced prosthesis, bioprinters and advanced 3D printers for medical application.

In addition, he is the founder and director of Concep3D Research and Development, where the design of products plays the main role. In this company, ideas are developed from scratch and utilize intellectual property, technical specifications, cost analysis and fabrication of pre-series products to support the release of new products into the market.

Summary

PREFACE	5
ABOUT THE AUTHOR	8
SUMMARY	10
1 INTRODUCTION	1
2 TECHNOLOGIES AND PIONEERS	9
2.1 FUSED FILAMENT FABRICATION (FFF) TECHNOLOGIES	12
2.2 TECHNOLOGIES BASED ON LIQUID POLYMERS	30
2.2.1 STEREOLITHOGRAPHY APPARATUS- SLA	31
2.2.2 MASK photopolymerization STEREOLITHOGRAPHY - MPSL	37
2.2.3 INKJET PRINT – IJP	42
2.3 TECHNOLOGIES BASED ON LAMINATED SOLIDS	45
2.4 TECHNOLOGIES BASED ON DISCRETE PARTICLES (POWDER)	52
2.4.1 SELECTIVE LASER SINTERING – SLS	52
2.4.2 3D PRINTING (3DP) - ZPRINTER	58
2.5 MAIN TECHNNICAL COMPARISON BETWEEN TECHNOLOGIES	61
3 THE RISE OF INDUSTRY 4.0	65
3.1 BREAKING PRODUCT DEVELOPMENT PARADIGMS	70
3.2 COLLABORATIVE DEVELOPMENT AND OPEN-SOURCE: ACCELERATORS FOR INDUSTRY 4.0	75
3.3 KEY 3D PRINTING APPLICATIONS FOR INDUSTRY 4.0	77
3.3.1 PROTOTYPE MANUFACTURING	78

3.3.2	SMALL AND MEDIUM SCALE PRODUCTION	79
3.3.3	HYBRID PRODUCTION IN LARGE SCALE	81
3.3.4	DISTRIBUTED PRODUCTION	84
3.3.5	RISKS AND LIABILITY	86

4 APPLICATIONS — 88

4.1	DESIGN AND ARCHITECTURE	88
4.2	MEDICAL AND HEALTH CARE	91
4.3	ENGINEERING DESIGN	97
4.4	MANUFACTURING AND PRODUCTION	102
4.5	RAPID TOOLING FOR INDIRECT MANUFACTURING	108
4.6	RAPID TOOLING FOR DIRECT TOOLING	114
4.7	WHAT IS THE MOST SUITABLE 3D PRINTER FOR EACH STAGE OF PRODUCT DEVELOPMENT?	116
4.8	DECISION MATRIX TO SELECT 3D PRINTERS AND APPLICATIONS	122

5 NEWS, MYTHS AND TENDENCIES — 125

5.1	MYTHS ABOUT 3D PRINTERS	125
5.1.1	CAN 3D PRINTER FABRICATE ANYTHING IN ANY SHAPE?	126
5.1.2	WILL 3D PRINTER REPLACE CONVENTIONAL PROCESSES?	131
5.1.3	CAN LOW COST 3D PRINTERS FABRICATE WEAPONS?	132
5.2	RESEARCHES AND INNOVATION	136
5.2.1	BIOPRINTING OR TISSUE 3D PRINTING	140
5.2.2	SLS WITH MULTIPLE MATERIAL	141
5.2.3	3D PRINTERS IN CIVIL ENGINEERING	143
5.2.4	SIMULTANEOUS DEPOSITION AND POLYMERIZATION – SDP	148
5.2.5	GENERATIVE DESIGN AND TOPOLOGICAL OPTIMIZATION	150
5.2.6	ADDITIVE MANUFACTURING WITH MULTIPLE DEGREE OF FREEDOM	152
5.2.7	MULTIPLE COLORS FFF	153

5.3 TENDENCIES _____ 156

1 Introduction

Over the years, due to consumer demands, companies have been driven to increase product complexity. On the other hand, to remain competitive, development deadlines had to be shortened.

In order to attend these new needs which were imposed by the market, several technologies were developed in order to innovate manufacturing systems, differing from machining, molding and forming, and others conventional fabrication processes.

Among these technologies, we can highlight the one which add material to manufacture objects layer-by-layer. This fundamental concept of fabricating objects has been spread over the years by various names. These names include Freeform Form Manufacturing, Rapid Prototyping, Layer Manufacturing, and 3DPrinting. However, due to the great evolution of these processes, there is a big effort worldwide to standardize these names for Additive Manufacturing. Therefore, ISO/ASTM 52900-15 established Additive manufacturing as the right terminology.

Although the concept of manufacturing layer-by-layer objects is old, the development of 3D printers began in the 1980s. At this time, these technologies were primarily focused in building prototypes quickly without molds,tooling or even material removal, as in mills and lathes. For this reason, the classification of these manufacturing processes as Rapid Prototyping had its most widespread diffusion in the industrial environment.

However, as these technologies mature, the number of applications has been gradually expanding into health, product development and production (Gibson 2005; Liou 2007; Volpato 2007; Cunico 2011; Cunico 2013).

In general, all 3D printing or additive manufacturing technologies have as their basic operating principle the generation of three-dimensional (3D) objects through the process of adding layer by layer material.

In Figure 1, a simplified schematic of this process is presented, indicating the main stages since the creation of the 3D computer model until the finalization of the part fabrication.

In this schematic, we present the five main steps related to the object manufacturing process in 3D printers: 1) 3D computer modeling; 2) Generation of STL mesh model; 3) Layer generation and manufacturing planning; 4) Layer-by-layer object fabrication; 5) Post processing and finishing.

The generation of 3D computer models is usually performed in design programs or computer aided design (CAD). However, layering and manufacturing planning typically uses small files which are called mesh, triangular, or faceted models. Therefore, 3D printing an object requires the 3D model to be converted from native files to STL (STereoLithography) format.

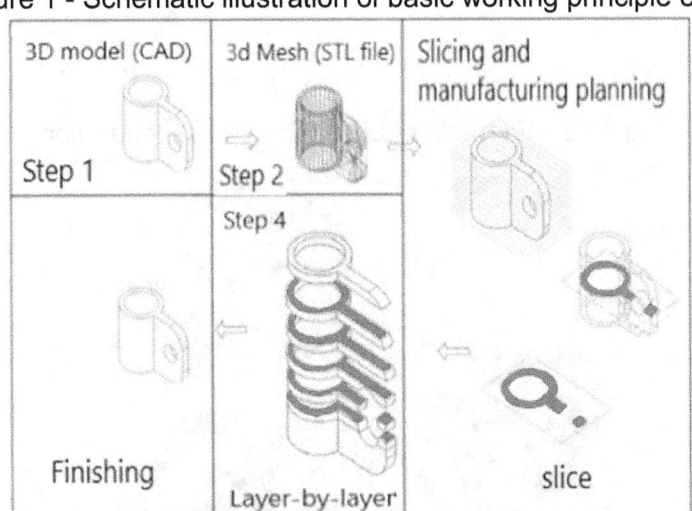

Figure 1 - Schematic illustration of basic working principle of 3D

On the other hand, in addition to the fact that these models provide constraints on geometric variation much higher than native CAD programs, they often have design defects, such as a lack of a face or triangle.

Additionally, as these model bases the construction of objects on triangles, the designer needs to specify the mesh size and scale appropriately.

Otherwise, the object will have shape implications, such as discontinuous surfaces. In Figure 2, an example of mesh model generation (STL) problems is presented, as well as the result of this problem in the manufacture of the final object.

Figure 2 - Example of unsuitable mesh generation

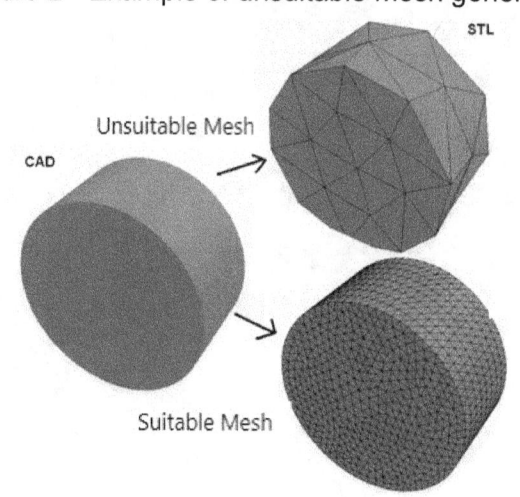

As a consequence, in many cases, performing finishing process to use objects that are visual in character is highly required (PRINZ, ATWOOD ET AL. 1997).

For example, fusion and deposition-based printers typically have deposited filament marks, as shown in Figure 3. These marks imply on the need for manual finishing at the end of object manufacture.

Figure 3 - Example of surface condition of FFF 3D printing part

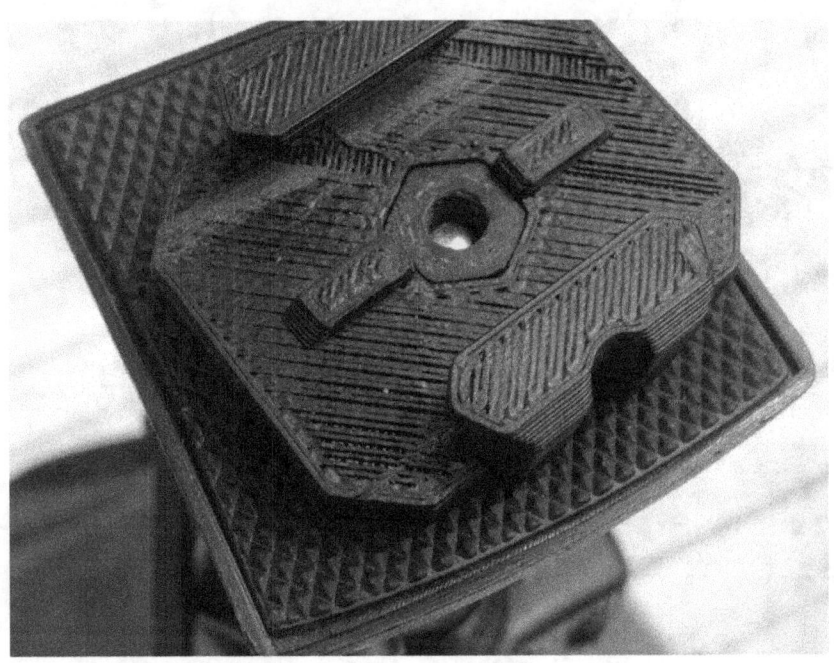

Similarly, another feature also found on objects made by 3D printers (additive manufacturing) is the incidence of layer marks and the occurrence of steps caused by high slopes, as illustrated in Figure 4. This feature is also known as the "stair effect".

Figure 4 - Example of Stair effect caused by inclining surfaces

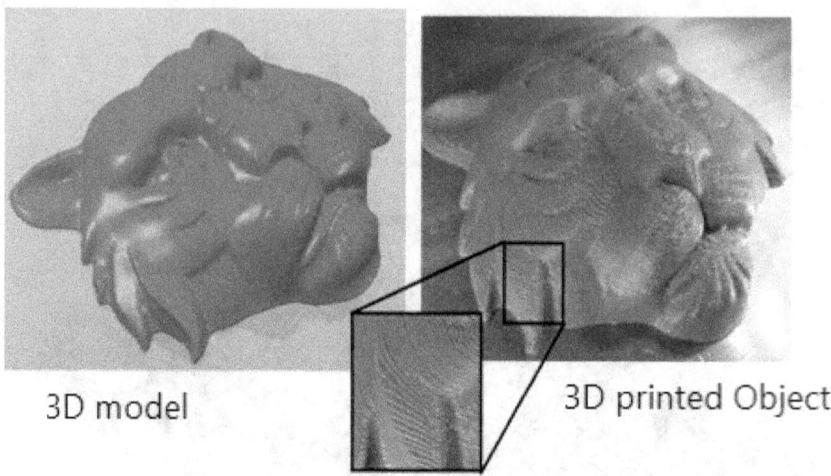

In this figure, an object manufactured in a fusion and deposition-based 3D printer is presented where the marks of manufacturing steps can be highlighted. In this case, it is necessary to use coating paints to achieve a suitable surface finish.

Despite the many constraints encountered by these processes, 3D printers have a differentiated manufacturing concept, allowing for more complex geometries than conventional processes, boosting their use in diverse areas such as health, design and engineering.

Figure 5 - Example of C3D Fusion 3000

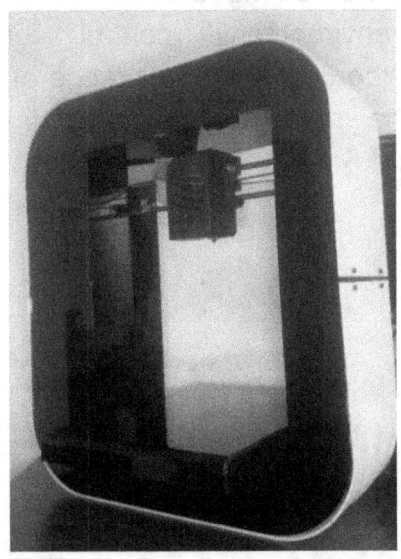

It is also possible to see trend of 3D printers popularization due to the launch of low cost products. Among these products, it should be noted that the process has a common concept, most of these printers are based on fusion and deposition of solid material.

In this low-cost segment, It is possible to highlight the top 10 companies: Makerbot, Ultimaker, BNCNC, Zmorph, 3D Systems, Stratasys, FormLabs, Siterize, Prusa and AnyCube.

Regarding the Brazilian national market, it can be seen that despite the great growth in the commercialization of this equipment, we highlight 5 main manufacturers (Sethi3D, 3D cloner, Cliever, Estela and Concep3D) in addition to some distributors of foreign technologies.

Among manufacturers, Concep3D introduces pioneering equipment among low-cost 3D printers, because dedicated computer-aided manufacturing (CAM) software and manufacturing environment control enable greater performance and the use of a wider range of materials (such as Ceramics). The image of the one of these products is presented in Figure 5

2 Technologies and Pioneers

The 3D printers development began significantly in the 1980s, where several proposals and patents were filled in addition to an extremely high volume of researches and development of new technologies. These efforts make 3D printers what they represent today.

Among the main milestones found in the evolution of 3D printers over the years, we highlight several concepts for the construction of tools from corded or laminated layers until the 60s.

From this date on, several proposals for 3D printers based on solidification of photocurable liquids have been proposed and patented (Stereolithography - SLA). Where Hull's patent in 1984 implied on founding of 3D System, one of the world leaders in the segment.

Another important milestone is the 1989 Crump patent that originated Stratasys and one of the most widespread technologies in the world: fusion and deposition modeling (FDM) or Fused Filament Fabrication (FFF).

Other technologies such as selective laser sintering (SLS) and laminated object modeling (LOM) also began in the 1980s and 1990s, where Deckard and Archella patents are milestones for SLS. For LOM, the biggest milestone was related to the filing of Feygin and Kinzie patents, which led to the founding of Helisys and Kira.

A chronological diagram of the major historical events in the evolution of material addition-based manufacturing technologies is presented in Figure 6.

In this figure, it can be observed that the birth of the concept of additive manufacturing had its cradle in the middle of mold making processes.

Over the years, the number of variants, configurations and applications of these technologies has been steadily increasing. However, stereolithography, FDM, LOM and SLS can be highlighted as the pioneering technologies that rose the 4th industrial revolution.

In order to facilitate the understanding of the 3D printing operation, several classification have been developed. One of them is based on the type of material used and processing used for the construction of objects.

In this type of classification, the following stand out:

- Fusion and deposition technologies (extrusion)

- Technologies based on liquid polymers

- Laminated solids technologies

- Powder-based technologies

Figure 6 - Time line of 3D printing Technologies

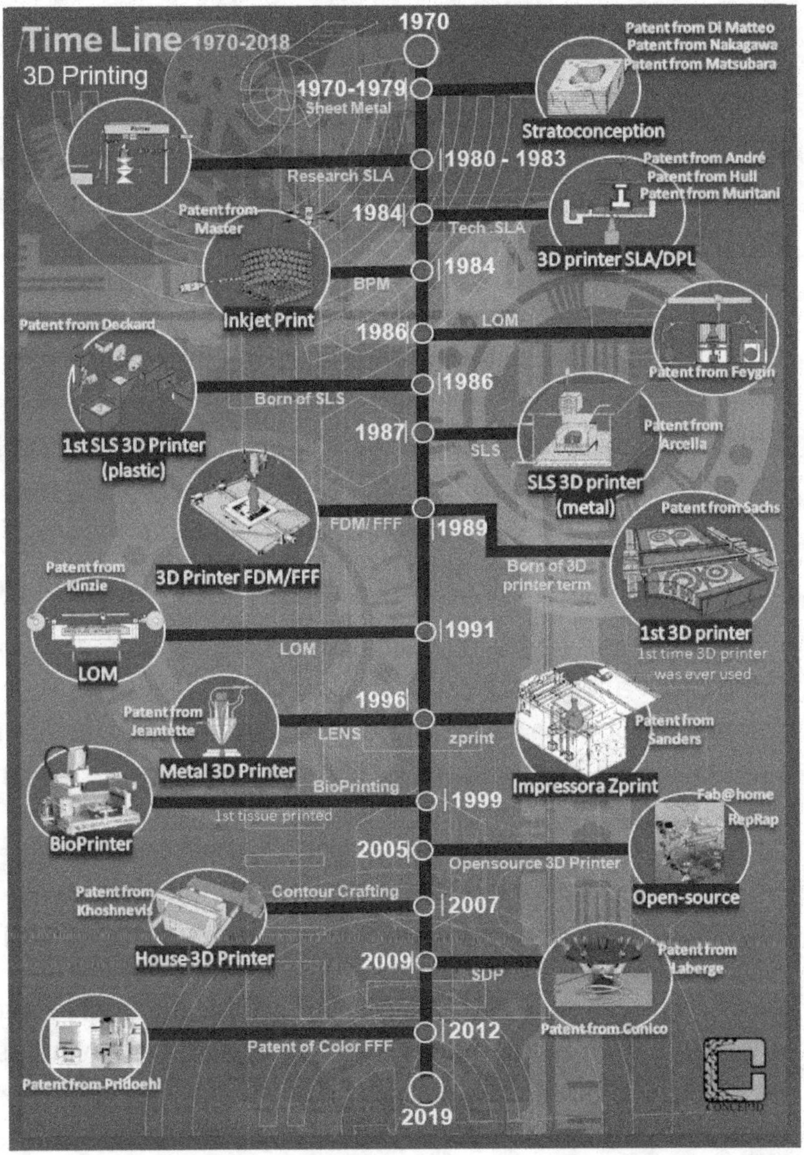

2.1 Fused Filament Fabrication (FFF) Technologies

Another classification for Additive Manufacturing (AM) technologies refers to the use of fused material. In this type of process, the principle of operation is the deposition of material, usually thermoplastic, through an extruder head.

In Figure 7, a schematic representation of the FFF or FDM process can be seen, where a filament of thermoplastic material is moved into the feed roller action blending chamber (normally driven by stepper motors). Around this chamber, thermal resistors are positioned to raise the material temperature to values above the softening temperature of the plastic (glass transition point).

In order to build the layer profile, this head moves along the x and y axes by adding material filaments. After the completion of each layer, the deposition platform moves in z direction for the purpose of building the next layer, repeating this procedure until the completion of the part .

Among the main commercial technologies using this principle of operation is Fused Deposition Modeling (FDM).

This technology, which is one of the most widespread in the market, has been developed by Stratasys, having as its initial milestone the patent filing of the technology by Stratasys founder (CRUMP 1989). This technology is one of the pioneers in the field of additive manufacturing , being the basic concept for the low-cost 3D printers found today.

Figure 7 - Schematic representation of Fused Filament Fabrication

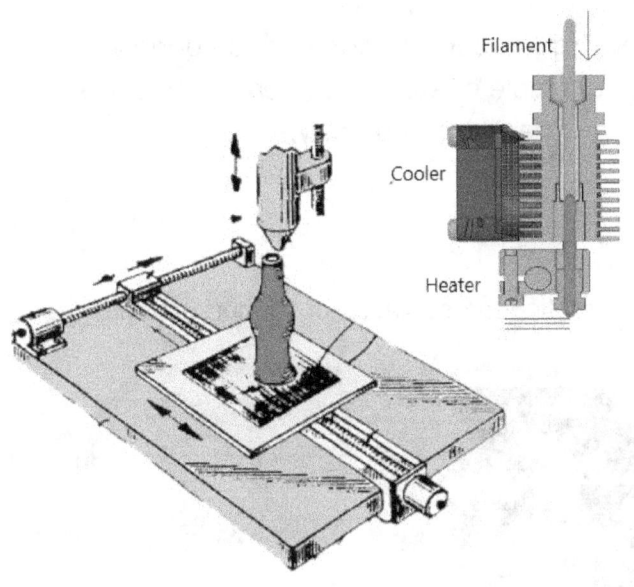

Because its operating concept is based on individual filament deposition, object manufacturing time is long, while other technologies such as the old 3D Systems Thermojet product provide simultaneous deposition through multiple deposition nozzles. In contrast, the material used by the former Thermojet is wax-based.

Regarding the basic processes involved in 3D printers, the following can be highlighted: material loading; material blending, applying pressure to propel material through the nozzle; extrusion; plot according to trajectory predefined by numerical commands; adhesion between building material; addition of support structure to allow the construction of negative and complex geometries.

Regarding the construction of such geometries, Figure 8 shows one example of support material which was used for the fabrication of negative geometry.

Figure 8 - Schematic representation of support material

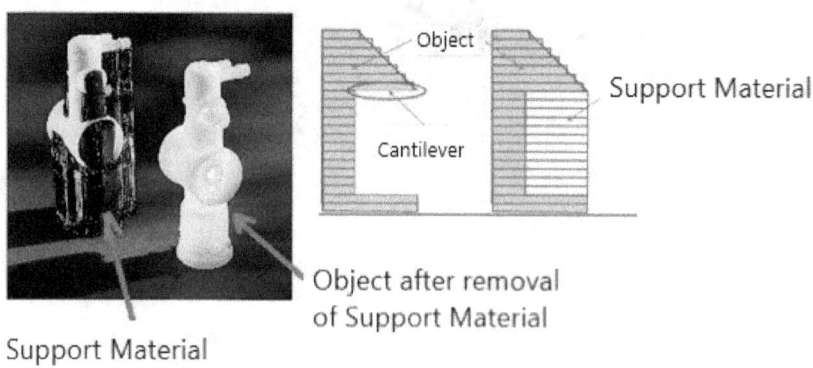

In this case, it is also necessary to remove this support material after the object is printed. This situation can often cause the object break or distort. For this reason, some companies tend to use water soluble support material, however the cost of this type of material is very high.

However, the use of support material implies a significant increase in manufacturing time and cost, as well as material waste.

In contrast, other branches of FFF varies according other material types and other principles of material solidification, such as: "ContourCraft", based on ceramics, Bioplotter, based on gel formation, chocolate printer, and FAB @ HOME, based on silicone.

Among the basic strategies of trajectory generation are the scanning and contouring, as can be seen in Figure 9. Usually, the raster path is the most used for the internal filling of the object, while the contour is used for the delimitation. of outer and inner silhouettes of the cross section of the object.

Figure 9 - Schematic representation of the basic layer profile for FFF 3D printers

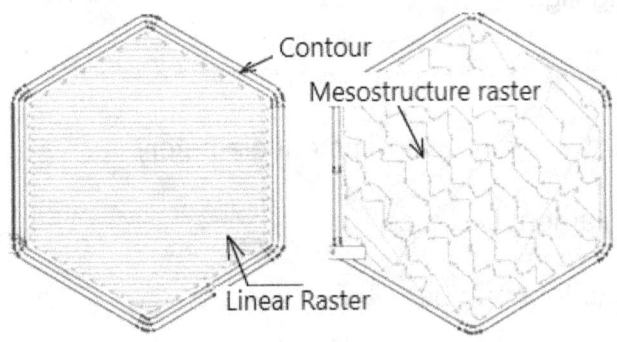

Key fill raster parameters include: filament distance, fill raster angle, and inter-layer raster rotation angle. However, the fill raster strategy may vary so that differentiated lattices, mesostructures or finishes can be obtained.

It should be noted that the form of infill directly affects the manufacturing time of the object as well as its resistance. However, currently the use of different forms of filling is becoming a common practice, as it can also be used for aesthetic characteristics.

An example of this type of application can be seen in Figure 10, where several types of fill are presented for the same object, besides being exposed the custom option of aesthetic fill.

Through this type of approach, objects can be manufactured with variable strength or even topologically optimized stiffness. It reduces cost and weight in addition to increasing strength.

Among the main control parameters of this process, we can highlight:

- Thickness

- Extrusion nozzle (filament diameter)

- Filament Distance

- Contour number

- Fill Density

- Fill Pattern Type

- Extrusion Speed

- Travel Speed

• Extrusion Temperature

• Extrusion Material

Figure 10 - Representation of type of infill

It is noteworthy that for each type of material has recommended parameters , such as extrusion temperature, which will imply on better finish objects.

Extrusion speed, displacement velocity and construction thickness are correlated variables. Thus, for low extrusion velocity values, the lack of deposited material can be observed when the high displacement velocity. On the other hand, high extrusion speeds imply on excessive material accumulation when the travel speed is low.

The thickness of construction affects the other manufacturing parameters, where large thicknesses can lead to lack of adhesion between layers and lack of deposition material when the extrusion speed is low and the travel speed is high.

In either case, the manufacture of the object is impaired, or even unfeasible. Thus, the choice of these manufacturing parameters must be made carefully so that objects are manufactured with quality.

Another point that is also important to be noticed is the distance between filaments. This parameter is fundamental for the object manufacture to occur suitably. If the distance is too small (less than the filament width), excessive material build-up can occur, while too large values will result in an open object and even no boundary between contour and fill.

The infill density, on the other hand, indicates the amount of material which will fulfill the inner part of objet as can be seen in Figure 11.In this case we can also determine the infill pattern.

In general, these parameters are also used for support material and raft configuration. Additionally, for determination of support material, the surface angle of the object must be identified so that the geometries requiring support might be identified.

Figure 11 - Illustration of infill as a function of density

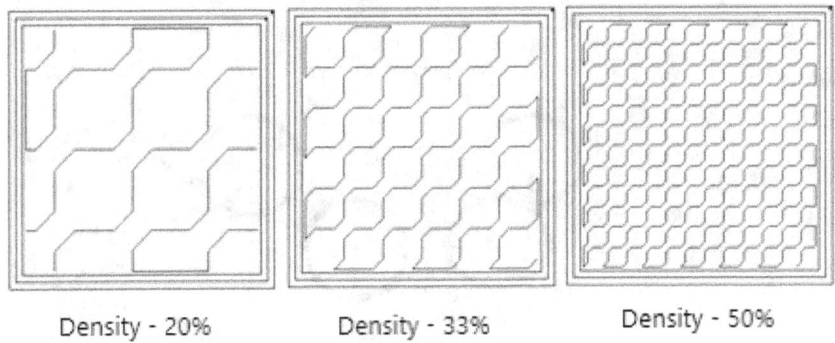

Density - 20%   Density - 33%   Density - 50%

In general lines, this surface angle is correlated to the need for support of one layer to rest on the previous layer. Thus, the smaller is the surface angle, the smaller is the support material. And consequently, in cases where there is a lack of support, it is necessary to use support, as observed in Figure 12.

It is also important to highlight that generation of support material affects several aspects of 3D printing, in addition to cost, printing time and material waste.

For that reason, several studies have been developed in order to optimize generation of support material.

Figure 12 - Representation of effect of surface angle on need for support material

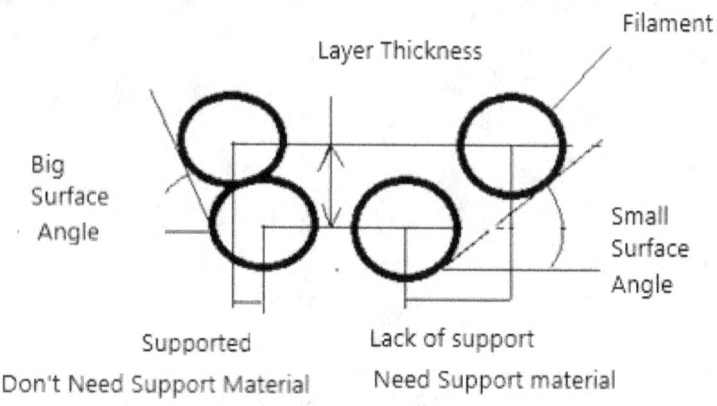

Figure 13 presents a comparison between default support material generator of slicers and optimized support material generated by Meshmixer. In this figure, we can see: a) optimized support material; b) support material with small angles; c) support material with large angles.

It is possible to see that default strategy resulted in 28.6% of waste material, in contrast with 6.6% of waste material which was generated by optimized strategy. It is also important to note that the fabrication time also reduced, making this approach even more attractive for the productivity point of view.

In addition, optimized support material also reduce risks of jeopardizing the object, whereas there is less support material in contact with the object to remove.

Figure 13 - Comparison between support material generated by slicer and optimized support material

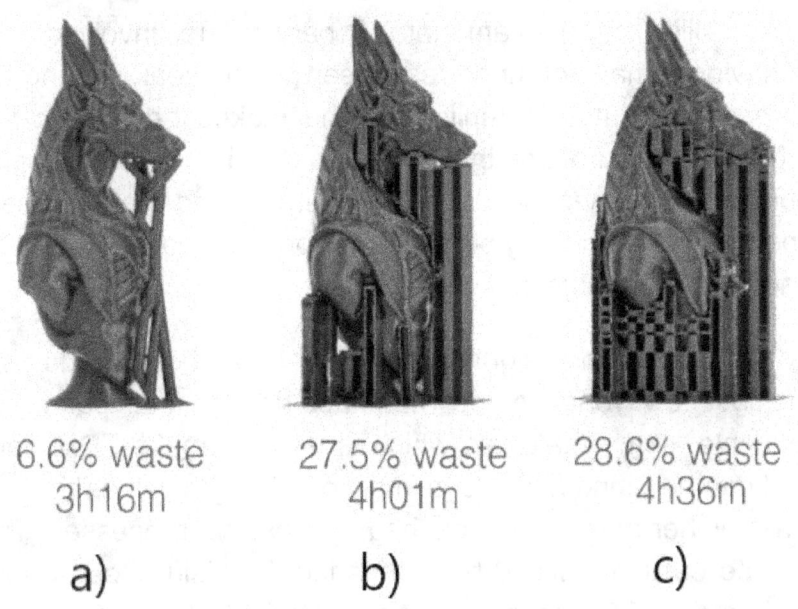

6.6% waste
3h16m

a)

27.5% waste
4h01m

b)

28.6% waste
4h36m

c)

This evidences the importance of design support material and plan the direction of fabrication during the concept and detail design stages.

In addition to the numerical command generated for deposition paths, the main variables that interfere in this process are: liquefaction chamber temperature, deposition room temperature, feed roller speed, traction gear drift deposition height, deposition speed and material thermomechanical properties .

Thus, we can see that, in spite this type of 3D printer is the simpler among 3D printers, the complexity of operation and fantastic results differing from 3D printer to 3D printers according to user (operator). Whereas the user that does not master to set parameters will perform a poor print.

Allied to the amount of parameters involved, the behavior of many of these responses is not linear, making the process even more complicated. This makes it difficult to find process configurations that covers to small, large, or complex objects simultaneously. With this, it is needed that we use a specific manufacturing strategy and process configuration for each particular object.

Despite this, significant progress has been made over the last few years in the manufacturing of final parts. For example, some materials which is used to make FDM parts and can be found nowadays imply on strength values that are even higher than those obtained by classical processes. This can be easily identified by comparing the yield stress values of Stratasys FDM thermoplastics material to those of molded materials. While the material values for FDM remain between 22 and 71 MPa, the injection mold equivalents are between 20 and 60 MPa (REDEYE 2011).

Another feature that should also be emphasized by this process is the precision in the construction of parts. For current equipment which is used for industrial applications, the minimum fabrication thickness (layer height) can be up to 0.127 mm. In addition, the precision along the layer (xy) is described as ± 0.127mm (GIBSON, ROSEN ET AL. 2010; REDEYE 2011; STRATASYS 2011).

Although this process is among the cheapest widespread in the industry, low finishing, low density and long manufacturing times are among the main limitations of using FDM technologies.

Due to the surface finish limitations, surface treatment processes are often performed on objects in order to obtain a better aesthetic and dimensional aspect, such as blasting, sanding, varnishing, among others.

Figure 14 - Comparison between object fabricated with layer thickness of 0.35mm before (a) and after (b) attack of vaporized solvent

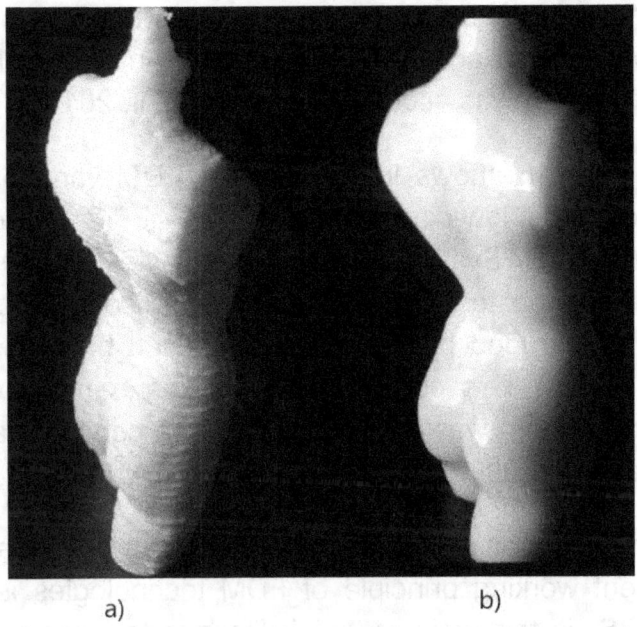

a)   b)

One example of thermo chemical treatment is shown in Figure 14, where the comparison between an object recently fabricated by FFF 3D printing and the same object after finishing process is presented.

In this example, the finishing is done through a process called "Smoothing", with Concep3D being the main provider of this type of service and equipment that provides this quality of finishing in Brazil.

Today you can find a number of low-cost variations of FFF/FDM equipment. This is a result of a hard work of projects that intend to popularize such technologies. We can highlight: Fab @ Home, which is based on its development at Cornell University; RepRap which was initially developed at the University of Bath. These projects generated companies that were acquired by big players, such as the Bitsfrombytes which was later acquired 3D System in mid-2010 and Makerbot, which was acquired by Stratasys in 2015.

Figure 15 shows images of two FDM variations that have pioneered low-cost products, a) FAB @ HOME and b) RapMan (3D Systems –Bits From Bytes). In the first example, the paste materials (such as silicone, chocolate, cheese) is extruded by a system composed by syringe and linear actuator. This concept was not brand new because deposition system was just a variation of a concept patented by Stratasys in 2008 (BATCHELDER 2008).

In the second conception (also patented by Stratasys), the current working principle of FDM technologies is used, using ABS as the main material. Despite the low purchase cost of these equipments, less than $20,000, they provide extremely low finishing and mechanical strength. Therefore it is often used at prototype level (GIBSON, ROSEN ET AL. 2010).

Figure 15 - Example of three of the first pioneers in open-source 3D printers

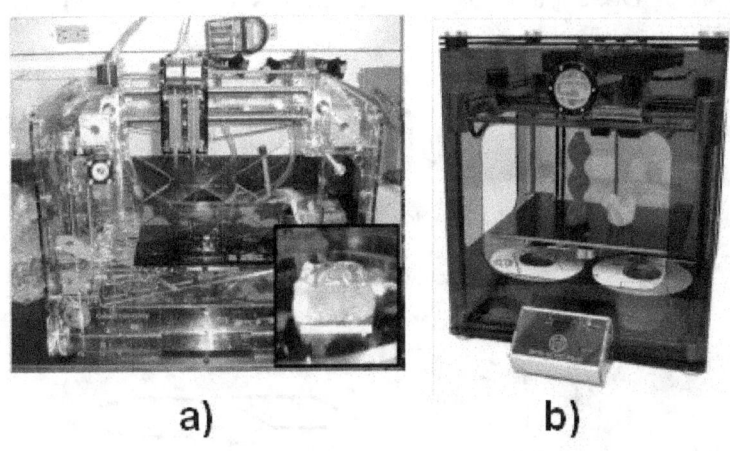

a)　　　　　　　　　b)

Nowadays, this technology has become popular due to the expiration of Stratasys patent in addition to initiative to popularize additive manufacturing technologies called RepRap.

From this initiative, several low-cost hobbyist variants were developed where the layout of positioning system and extruder vary in accordance with the product popularity.

For example, we can express the layout of 3D printers positioning system through kinematic schematics, as presented in Figure 16. In this figure, we can identify as basic elements: Translation elements, rotation elements, building platform, and extruder. It is also important to note that the layout nomenclature is defined by the motion of extruder follow by the motion of platform, as XY-Z. In this example, extruder moves in xy directions while platform moves along z direction.

Figure 16 - Description of main elements of positioning system schematic

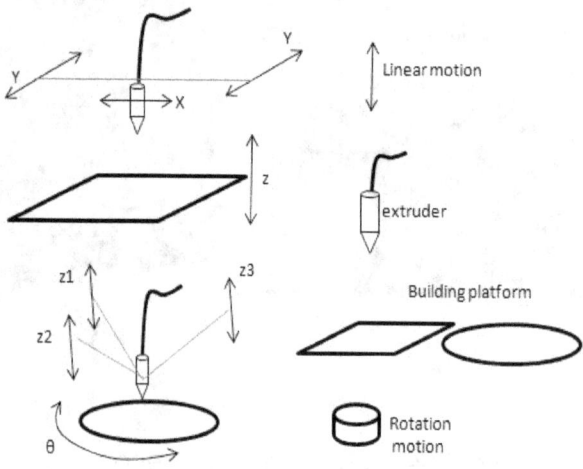

The most famous concepts being presented in Figure 17 where the schematic of positioning system and example of product which use such type of systems have been presented.

Several equipment configurations are presented in this figure, where the construction platform and deposition head movement orientation are changed to result in various product variants.

Figure 17 - Diagram with the main 3D printer layout and their respective popularity

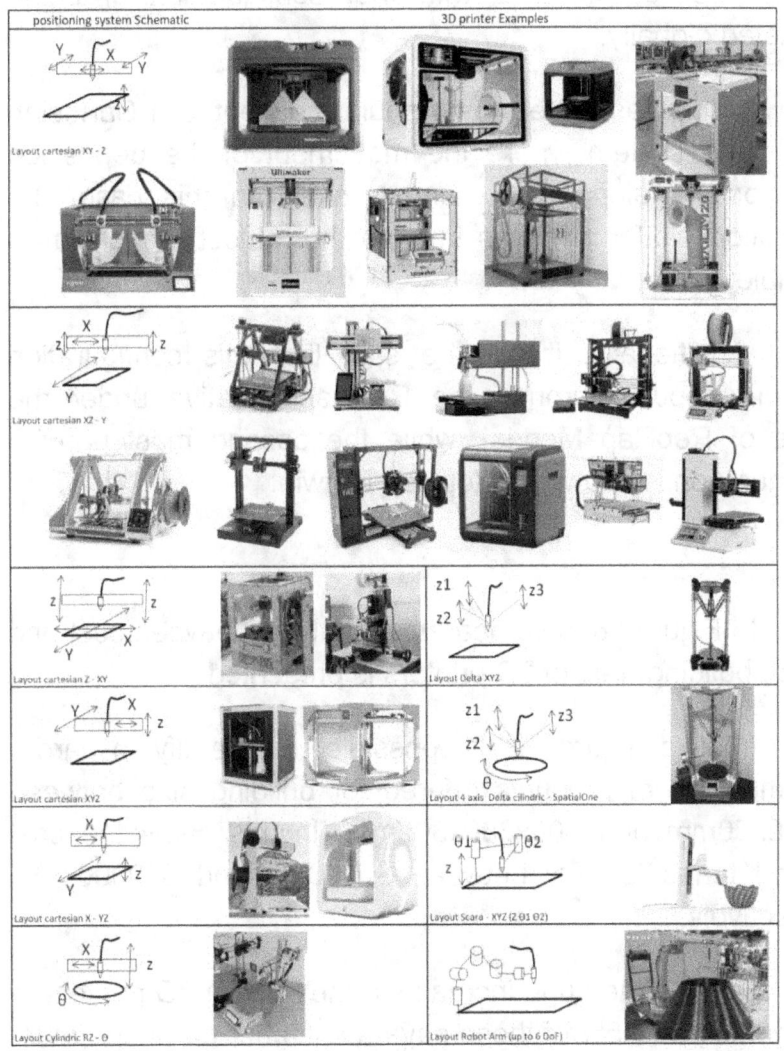

It can also be commented that this figure presents a popularity index of the different configurations in the market. This indicates that the higher the popularity rating, the greater the number of products with that configuration of being marketed globally.

In this case, the most popular concept is a Cartesian positioning system where the movement of the deposition head occurs along the x and z axes. In this case, the deposition platform moves in y so that object fabrication is possible.

Additionally, it can be said that this configuration became popular through the RepRap initiative under the name of RepRap Mendel, while the second most popular configuration is known as RepRap Darwin.

In Figure 18, a comparative analysis between cost and size of building area of 3D printers is presented.

In this figure, it is possible to identify a larger accumulation of products offered for building size between 50x50x50mm and 300x300x300mm. In this case, the cost offered tends to vary between $ 250.00 and $10000.00 , respectively.

Despite the huge increase in the offer of 3D printers in the market, only 5% of these have a chamber for temperature control. It is important to know that temperature control in building chamber increases the quality of printing, in addition to avoiding warpening, layer delamination, and surface roughness.

It can also be noted that the incidence of LOM technology is also rare because its use is very restricted by paper and laminated plastics.

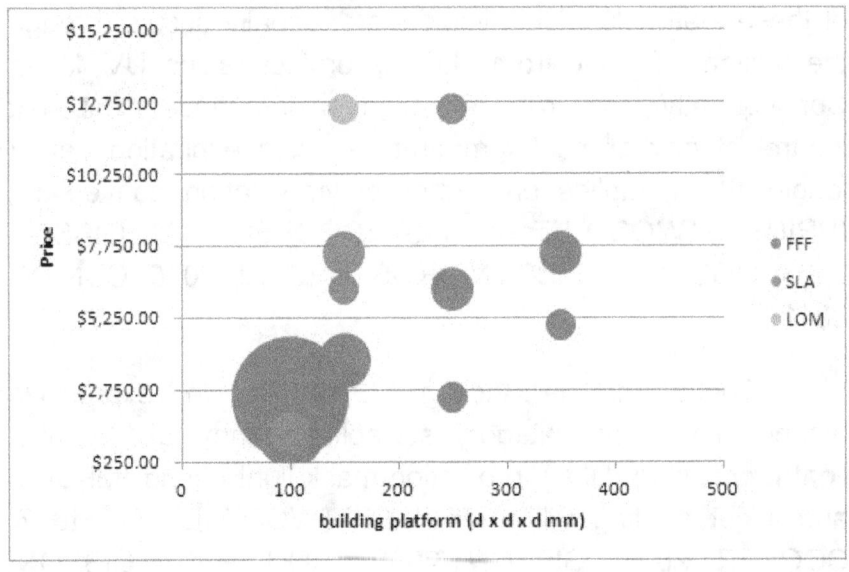

Figure 18 - Comparison diagram with cost vs type of 3D Printer vs building volume

On the other hand, despite the high level of resolution and finish, SLA technologies still cost more than FDM, even for equipment with a manufacturing size between 50 and 150 mm.

2.2 Technologies based on liquid polymers

With regards to technologies based on liquid polymers, these have the use of light curing materials as common characteristic. The most common based of these materials are polyurethanes, styrene, acrylates and epoxies. The cure of these materials, still in liquid state, usually occurs through the action of light from, LEDs, projectors or UV laser. Consequently, this requires high maintenance cost, big control of environment temperature, resin expiration date (longer times implies on higher polymerization conversion) (PRINZ, ATWOOD ET AL. 1997; COOPER 2001; GIBSON 2005; LIOU 2007; GIBSON , ROSEN ET AL 2010; CUNICO 2011).

These include the most well-known expressive technologies in this category: stereolithography (SLA), digital light processing (DLA), projection mask light curing (MPSLA) and inkjet printing (IJP). (PRINZ, ATWOOD ET AL. 1997; COOPER 2001; GIBSON 2005; LIOU 2007; GIBSON, ROSEN ET AL. 2010; CUNICO 2011)

2.2.1 *Stereolithography* Apparatus- SLA

Stereolithography Apparatus (SLA) is one of the most widespread rapid prototyping technology, initially marketed by 3D Systems.

Although the validation of its functional principle has firstly been published by Kodama (1981) and Herbert (1982), SLA technology was originally pioneered by the founder of 3D Systems only in 1984 (HULL 1986; PRINZ, ATWOOD ET AL. 1997). ; GIBSON 2005; GIBSON, ROSEN ET AL. 2010).

The functional principle of this process is the localized curing of photosensitive resin by UV laser beam which moves along the X and Y axes. This beam focuses on a resin-immersed container to construct the previously computationally calculated layer silhouette.

Upon completion of each layer, the material attached to immersed platform moves along the Z axis, allowing the beginning of construction of a new layer, as can be seen in the illustration of Figure 19.

This process is repeated until the workpiece is finished, when the platform emerges from the object, allowing the removal of unpolymerized resin from the object and increasing the resistance of the object in the oven .

Figure 19 - Generalized representation of bottom-up Stereolithography (SLA) process

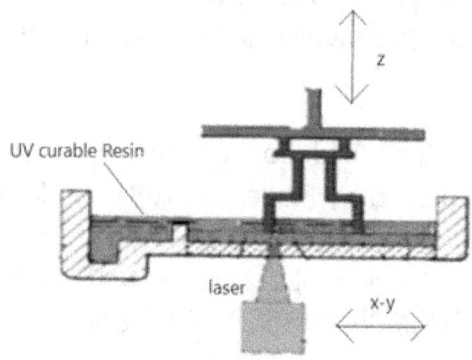

This process can also be observed in Figure 20, where an example of a skull being made in stereolithography (SLA) is presented. In this figure, the laser scanning step during the construction of the layer is presented, as well as the finishing of the part with support material, and final object obtained after removal of support material and post curing.

It is important to note that although the current design of SLA uses UV laser, the original patent of this process claims the possibility of using other sources that promote polymerization, such as electron beam (EB), radiation, high particle beam. energy, x-ray, UV beam and conventional UV light (HULL 1986; WU, ZHAO ET AL. 2001).

Figure 20 - Example of Top-down SLA fabricating an skull for sirurgical model

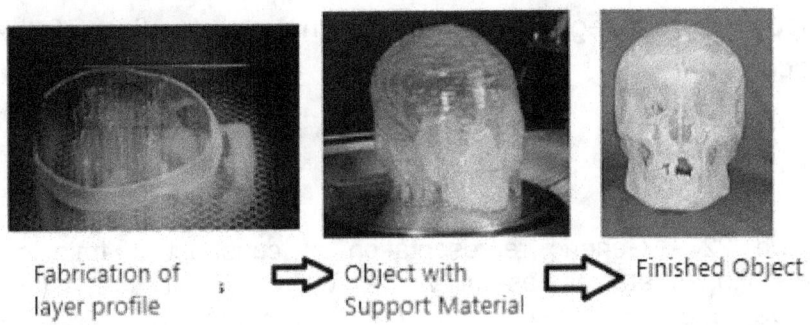

Fabrication of layer profile ⇒ Object with Support Material ⇒ Finished Object

Regarding the healing behavior generated along the laser displacement (ScanPattern), Figure 21 shows an illustration of the cross section of the filament generated by a laser displacement along a straight line (JACOBS, REID ET AL. 1992; COOPER 2001; GIBSON, ROSEN ET AL. 2010).

Figure 21 - Exampe of transversal cut of a polymerized straight line

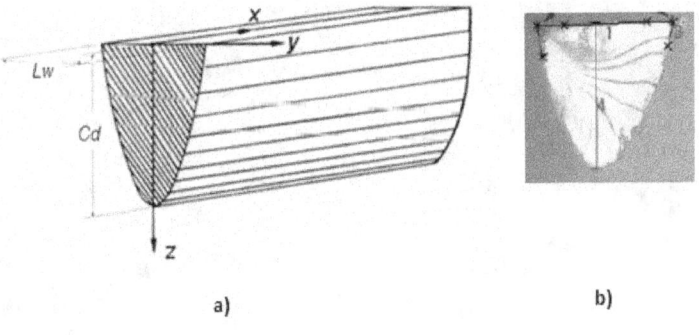

a)   b)

In this figure, the interaction between scanned filaments is also presented, indicating the existence of unpolymerized regions throughout the construction of the piece.

Figure 22 - Generic representation of construction trajectory generation strategies, as follows: a) weave; b) 3D mesh (star weave); c) contour; d) scanning, adapted from (JACOBS, REID ET AL. 1992; GIBSON, ROSEN ET AL. 2010)

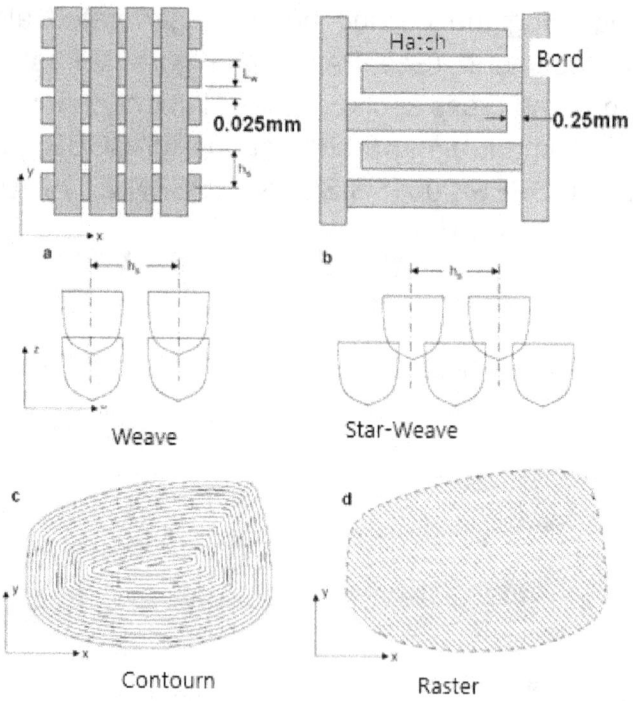

Besides the existence of these unpolymerized regions, the degree of polymerization of the newly fabricated part is 80 to 90%, implying on the completion of a post-processing in order to finalize polymerization and increase mechanical strength of object. At this stage, the finished object is placed and kept in an oven that has a UV light source and emits heat for 1 to 2 hours and the support material is removed at the end of the process (JACOBS, REID ET AL. 1992 ; COOPER 2001; LIOU 2007; GIBSON, ROSEN ET AL. 2010).

The main trajectory strategies used in this process are: sweep, contour, weave and three-dimensional mesh (star-weave), as can be seen in Figure 22.

The materials typically used by this process are based on urethanes, acrylates and epoxies, providing for end objects a mechanical strength suitable for manufacturing, various applications and assembly.

In addition, the mechanical strength of these parts can reach levels equivalent to those of plastic parts manufactured in conventional processes (28 to 78 MPa)

With regard to the accuracy of this technology, layer resolution values of up to 50μm are currently found, while for sweeping accuracy up to 25μm.

The construction speed of objects can reach up to 35m/s. However, there are specific cases, such as micro stereolithography, whose layer resolution (z) and scan accuracy (xy) values can reach 0.1 and 0.25μm respectively

(3DSYSTEMS 2010; GIBSON, ROSEN ET AL. 2010; 3DSYSTEMS 2011; 3DSYSTEMS 2011).

Figure 23 - Office environment stereolithography equipment - FormLabs where: a) SLA; b) washing equipment; c) UV Oven

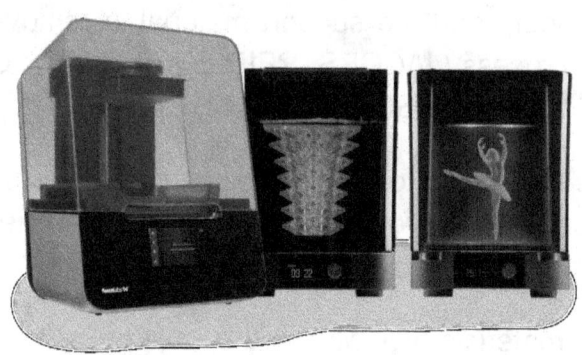

Nowaday, despite resin toxicity, there is a trend toward the release of SLA technologies for the office environment, where it is possible to obtain objects with greater accuracy than other popular processes.

Figure 23 shows a picture of one of the pioneers of stereolithography equipment in the home environment. This equipment is manufactured by Formlabs company, founded in 2013 in through KickStart project.

As an example, the main differences between this process and those based on domestic environment fusion and deposition can be seen in Figure 24, which shows the greater precision of this technology.

Figure 24 - Comparison between home environment equipment: FDM and SLA

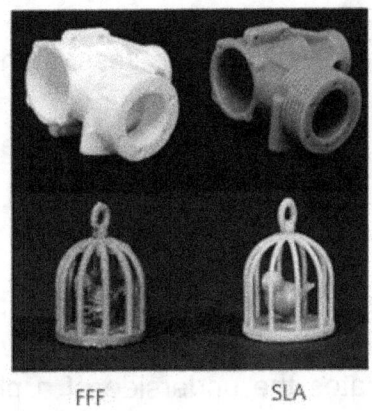

Although this technology is more accurate, the thermal resistance of objects obtained from this type of printer is low, where the softening temperature (glass transition temperature) is between 40 and 80°C. Thus, the type of application of this technology ends up being restricted to the environment where the object will be used.

2.2.2 Mask photopolymerization Stereolithography - MPSL

Another 3D printer using today's widespread photopolymerizable materials is Mask photo polymerization Stereolithography (MPSL), which started being called DLP

(Digital Light Processing) nowaday. This process, which is also known as photo solidification, began to develop in parallel with the development of SLA, however, beginning with the commercialization of equipment in 1991.

The operation of this process stands out from the others due to the layer creation mode, since each layer is generated in a single step (layer-wise), reducing the construction time of the part and increasing the homogeneity of the layers.

A schematic example of the operation of this technology can be seen in Figure 25, where a projector selectively illuminates the underside of a photocurable resin container with a cross-sectional image of a layer (PRINZ, ATWOOD ET AL. 1997; GIBSON 2005; Gibson, Rosen et al.

By the end of each layer, the construction platform moves upwards, allowing a new layer to be constructed and adhered to the previous layer.

An example of this can be seen when analyzing the manufacturing speed of EnvisionTec's Perfactory® Standard UV equipment. As this provides a manufacturing speed of 200 layers of 100 μm thickness per hour, the construction of a piece of volume equal to 2484 cm^3 would occur in 11 hours.

The construction of a part of the same proportions through an SLA process would be performed at the same time, but obtaining higher construction resolution. Therefore, a greater detail of the part can be obtained in this process (PRINZ, ATWOOD ET AL. 1997; GIBSON 2005; ENVISIONTEC 2009; ENVISIONTEC 2009; GIBSON, ROSEN ET AL. 2010; ENVISIONTEC 2011).

Figure 25 - Generic schematic of bottom-up SLA called DLP

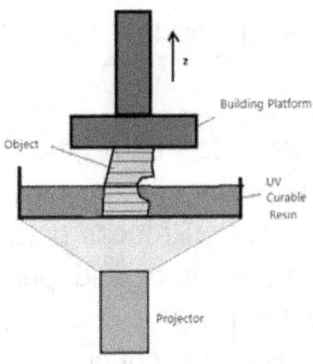

For the construction of each layer, UV light is projected through a mask or multimedia under a container filled with photo curable resin. Because the layer silhouette is projected entirely through the mask or multimedia equipment, material solidification occurs only in the layer format. At the end of each layer, a material support platform moves vertically (z axis) allowing the construction of the new layer. This process repeats until the part is finalized and the part can be removed from platform, cleaned and put on the oven for finishing the polymerization.

Although several groups have been continuously studying this process, currently only two companies are prominent in marketing. EnvisionTEC had its products launched from 2003, while the 3D System only in 2008, with V-Flash equipment (PRINZ, ATWOOD ET AL. 1997; GIBSON 2005; GIBSON, ROSEN ET AL. 2010).

Nowadays, several low-cost MPSLA are possible to be found in the market with the name DLP (Digital Light Projector). The price of this low cost SLA/DLP varies from $200 to $5000.00.

It is important to note that among the main weaknesses of these low cost SLA/DLP we can highlight: Resin quality and life-time, Process parameters, base quality and life-time, support material and part orientation, optical distortion correction. For those reasons, low cost SLA/DLP are extreme hard to use and configure.

On the other hand, the surface finishing and accuracy of those 3D printers are by far better than FFF 3D Printers, been largely used in dentistry industry (surgical models).

The mechanical properties of parts manufactured by this process can be up to 1.8 times the strength of conventional plastics (57MPa), such as ABS, yet still below the maximum SLA values (PRINZ, ATWOOD ET AL. 1997; GIBSON 2005; GIBSON, ROSEN ET AL 2010.; 3DSYSTEMS 2011; ENVISIONTEC 2011; ENVISIONTEC 2011; ENVISIONTEC 2011; ENVISIONTEC 2011; ENVISIONTEC 2011)

Another example of the manufacturing process and objects removed after construction is completed can be seen in Figure 26. This figure shows a photo of the equipment as well as an image representing the construction of the object from the customer's perspective and the objects after removal from the platform. .

Figure 26 - Example of Envisiontec bottom-up Stereolithography

In this case, support material removal, residual resin removal and cure completion are still required. This implies on complexity and issues about the use of this type of technology in the office environment.

Additionally, it can be emphasized that a "construction bed" is normally used that facilitates the removal of objects from the platform without damage or loss of finish. Otherwise, if the object is built directly onto the platform, the process of removing the platform would jeopardize the object's finish or even damage it.

2.2.3   Inkjet Print – IJP

With reference to the rapid prototyping technologies based on inkjet print (IJP), the working principle is to create layers by depositing droplets of material through the inkjet head.

This technology was first developed by Sanders Prototyping (SolidScape) in 1994. However, the use of inkjet in AM equipment was most significant only from 1999, when the design employed by Israeli company Objet Geometries (PRINZ, ATWOOD ET AL 1997; SANDERS, FORSYTH ET AL 1998; GOTHAIT 2000; GIBSON ROSEN ET AL 2010). In 2013, this company merged with Stratasys, the leading manufacturer of FDM processes.

In the initial concept, presented by Sanders Prototyping, a thermal inkjet head is used to selectively deposit drops of wax-based materials for cross-sectional construction of 3D objects along the x and y axes.

After completion of each layer, the construction platform shifts in z providing the construction of the next layer, and this process repeats until the completion of the part (PRINZ, ATWOOD ET AL. 1997; SANDERS, FORSYTH ET AL. 1998; Gibson, Rosen et al.

On the other side, the design initially adopted by Objet is based on piezoelectric inkjet head which is used to deposit droplets of liquid photo curable material onto a platform or substrate along the x-y axis.

Figure 27 - Schematic of polyjet printer from Stratasys

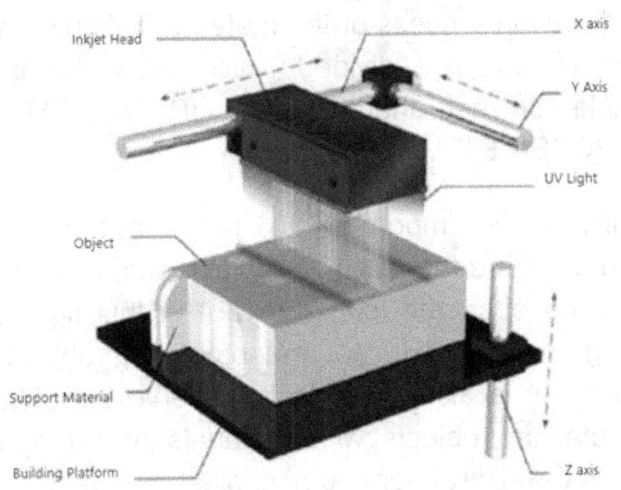

Simultaneously, these drops are solidified due to exposure to an extensive UV light source. Upon completion of each layer, the construction platform shifts in z providing the start of a new layer, as can be seen in Figure 27 (PRINZ, ATWOOD ET AL. 1997; GOTHAIT 2000; GIBSON, ROSEN ET AL. 2010).

Facing the Stratasys PolyJet line (formerly Objet), the company 3D Systems, through the Multi-jet line, adopted a differentiated solution from the conception initially developed by Objet. In the case of this design, instead of the material drops solidifying simultaneously, they are partially cured by flashes of UV light at the end of each layer (SCHMIDT 2004; MARGOLIN 2006).

The type of materials typically are usually based on urethane, acrylates, epoxies, thermoplastics and wax, which differ by the solidification process of the material. This allows these technologies to be divided into two groups: photopolymerizable IJP and thermoplastic IJP (VOLPATO 2007; GIBSON, ROSEN ET AL. 2010).

Additionally, it is important to note that some technologies also allow the fabrication of objects with multiple materials, as can be observed in Figure 28. In this figure, besides indicating the possibility of manufacturing objects with different colors, it is also possible to observe that it is possible to manufacture objects with materials of different properties, such as flexible rubber and rigid plastic.

Figure 28 - Example of objects fabricated with multiple material polyjet

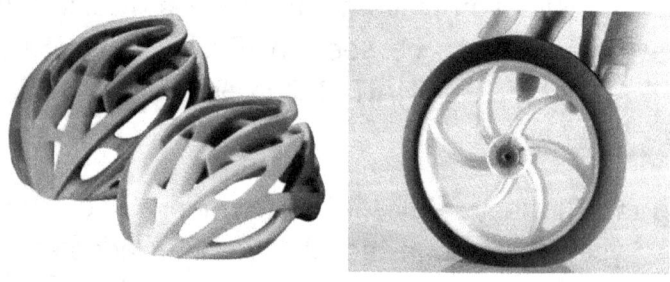

For example, EDEN equipment, developed by Stratasys (Objet), is an IJP type of light-curing materials, and

its inkjet head deposits material through 1536 individual nozzles arranged in line simultaneously in 15 μm layers. In this case, since UV curing occurs simultaneously with deposition, post-processing is not fully necessary, as in other liquid-based processes (SLA, DLP and MPSL).

Regarding the mechanical properties provided by these processes, the main tensile strength (yield strength) values found in IJP photopolymerizable technologies varies from 30 to 76 Mpa. On the other hand, accuracy and resolution of these technologies are around 600 dpi horizontal (xy), which is equivalent to 42 μm .

2.3 Technologies based on laminated solids

Another classification of 3D printers is based on the use of laminated solids. This AM technology was one of the first to be commercialized, starting in 1991. Initially, the Japanese company that developed this technology (Kira) called it the Laminated Object Modelling (LOM) process.

The principle of operation of this 3D printer is based on cutting and pasting sheets of paper or plastic. The generation of the layer profile is performed by CO_2 laser cutting long the xy axis. The layers are gradually stacked and glued through thermal resins.

Figure 29 - Schematic of LOM working principle

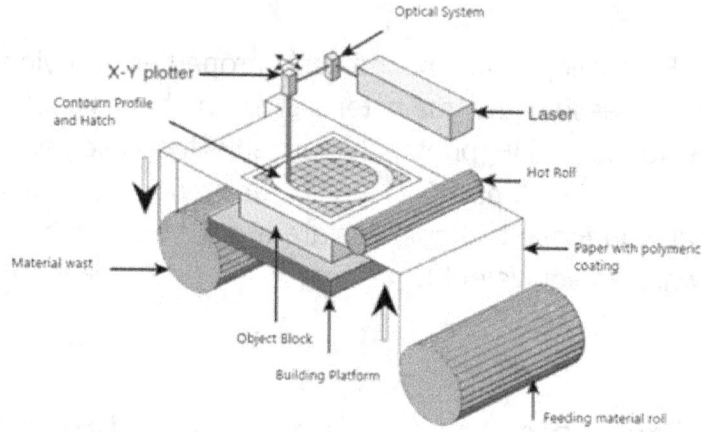

As we can see in Figure 29, after completion of each layer, a platform is moved on the z axis, allowing the beginning of a new layer, where the new layer is glued onto previous layer. For bonding between layers, a roll compresses the layer to be fabricated against the previous layers, providing adhesion through pressure and heat (PRINZ, ATWOOD ET AL. 1997; COOPER 2001; GIBSON 2005; GIBSON, ROSEN ET AL. 2010).

On the other hand, the basic principle of operation has varied over time, such as the use of cutting blade and inkjet head for the generation of colored parts, and the use of materials apart from paper.

Another sub-classification of these technologies can be identified according to the order of processing, such as Form-Then-Bond and Bond-Then-Form. In the case of Helisys LOM, KIRA and Solido, these are based on the glue-cut classification, using, respectively, roll paper, sheet paper, and polypropylene roll (PRINZ, ATWOOD ET AL. 1997; MORITA AND SUGIYAMA 2000; GIBSON 2005).

Figure 30 - Example of fabrication block (removal part) and fabricated object

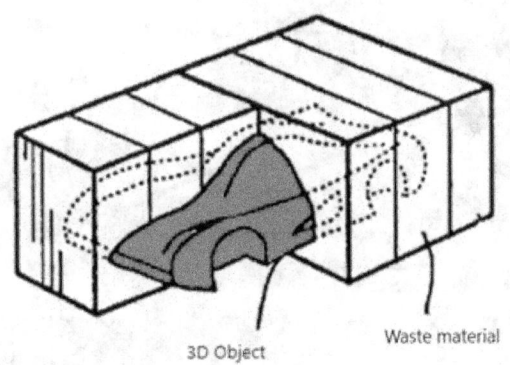

In the case of conventional LOM processes, a cellulose block is obtained at the end of the object construction, and a post-processing process is required to remove non-part areas as can be seen in Figure 30.

It is important to note that the non-part area is material waste. Therefore, fabrication of small and large objects with same height will consume the same amount of material.

It evidence one of the biggest disadvantages of this technology, material waste.

Nowaday, there has been an evolution of this technology, where the use of colors makes the process interesting especially in relation to design related applications. Examples of objects made by this type of equipment can be seen in Figure 31.

Figure 31 - Examples of objects manufactured by Mcor Iris equipment from the Irish company Mcor Technologies

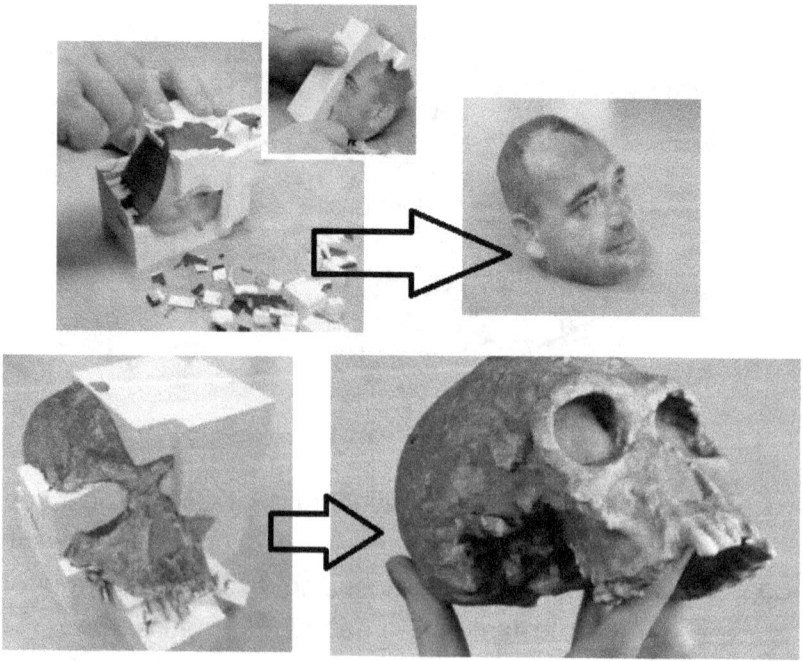

Despite the many benefits of this type of 3D printer, the final material is paper based and not water resistant. It can also be noted that the mechanical strength of this process is not high, as the manufacture of small geometries is barely possible because of the fact that removal of waste material is very aggressive. Thus, the possibility of damage to the part is high.

On the other hand, manufacturing time of objects with geometries that have large cross-sectional areas is reduced compared to other processes, since this type of printer does not have a filling routine in the making of layers.

It is also possible to note that despite low cost of raw material, this process imply on extensive material waste. For that reason, small objects are unsuitable to be produced in such technology.

On the other hand, among the main processes based on glue cutting are: Stratoconception; CAM-LEM and Offset Fabbing system, having respectively as base materials: wood, metal and ceramic plates (WEI, CHOI ET AL. 2001; HOUTMANN, DELEBECQUE ET AL. 2009; GIBSON, ROSEN ET AL. 2010; THABOUREY, BARLIER ET AL, 2010).

Figure 32 shows the schematic illustration of Stratoconception and CAM-LEM processes.

Figure 32 - Schematic illustration of Stratoconception and CAM-LEM process (GIBSON, ROSEN ET AL. 2010)

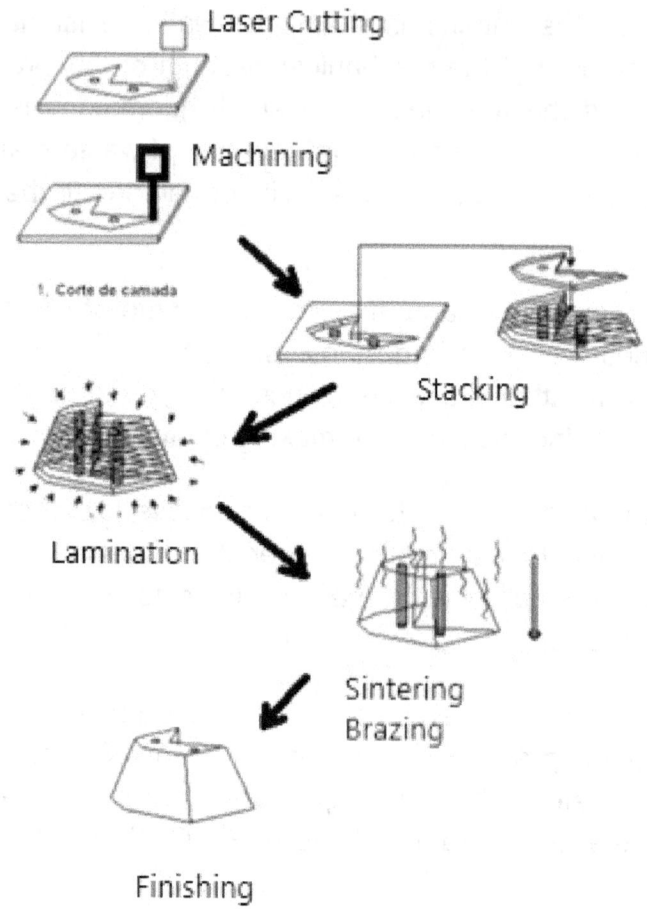

In this process, the layers of the 3D object are initially cut by laser cutting or CNC, and the stacking procedure of these layers is performed manually or automatically using clamping pins. Then, the object has to be brazed or bonded, where the fixing pins may or may not be removed. It is possible to note that the 3D object may be finished according to surface condition needs.

An example of this type of wood application can be seen in Figure 33, where the layers are produced by wood sheets and joined by glue and pins. In this case, lamination was performed by gluing and the binder removal phase was ignored.

Figure 33 - Ilustration of object fabricated by wood Stratoconception

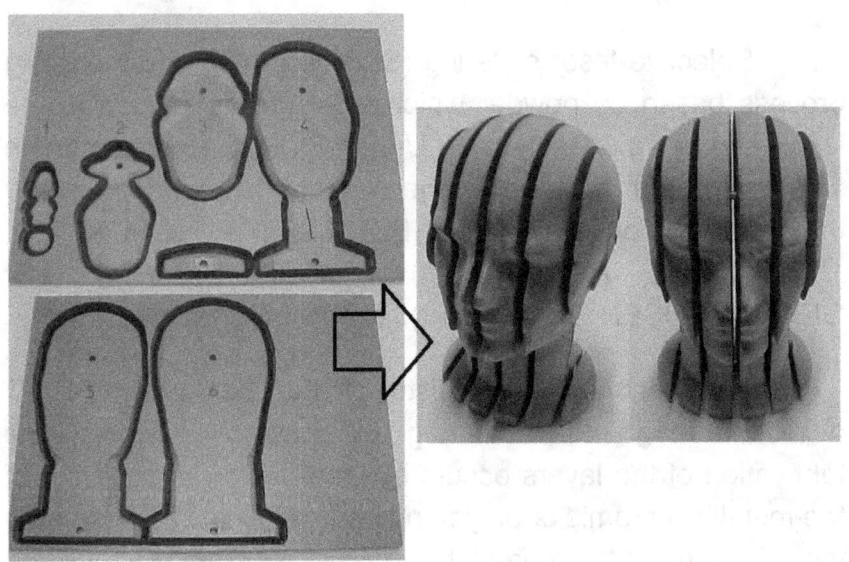

2.4 Technologies based on discrete particules (powder)

Another classification of AM technologies has its working principle based on discrete particle or powder based materials. These processes include Selective Laser Sintering (SLS), Selective Laser Melting (DSLM), Laser Engineered Net Shaping and 3D Printing.

2.4.1 Selective Laser Sintering – SLS

Selective laser sintering (SLS) is a type of 3D printing process based on powder which was first developed at the University of Texas at Austin, United States, and patented in 1989. However, their commercialization began in 1990 through the company DTM (BEAMAN AND DECKARD). 1990; PRINZ, ATWOOD et al. 1997; GIBSON 2005; GIBSON, ROSEN ET AL. 2010).

This technology consists of the fabrication of 3D objects through the sintering of material powder. The fabrication of the layers occurs by laser fusion or sintering of the metallic, ceramic or polymeric particle, which moves along the XY axes. After completion of each layer, the building platform moves on the z axis allowing material to be fed into the new layer, as can be seen in Figure 34(PRINZ, ATWOOD ET AL. 1997; COOPER 2001; GIBSON 2005; LIOU 2007; VOLPATO 2007; GIBSON, ROSEN ET AL. 2010).

Figure 34 - Schematic illustration of Selective Laser Sintering processes

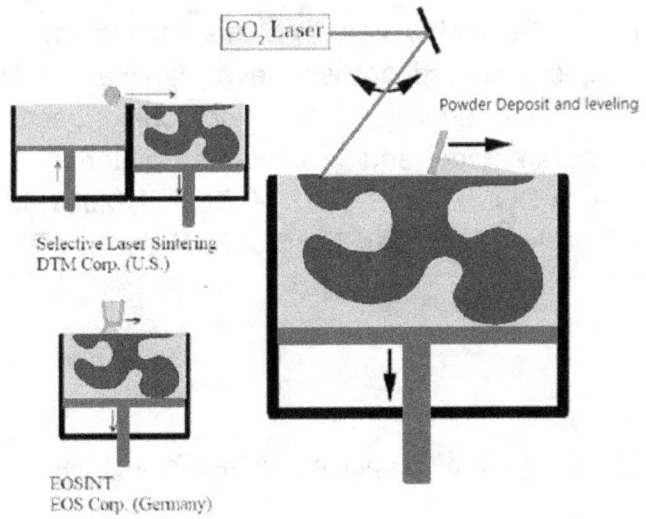

These steps are repeated until the end of the object, which in many cases requires a post-processing process to obtain mechanical strength. In these cases, the part that is also called the Green or Gray Part is infiltrated by binder material in addition to the burning of residual material (PRINZ, ATWOOD ET AL. 1997; COOPER 2001; GIBSON 2005; LIOU 2007; VOLPATO 2007; Gibson, Rosen et al.

In addition to being one of the first technologies to be commercialized, this category of 3D printers offers the possibility of manufacturing one of the largest varieties of material among AM processes. However, due to the large number of parameters involved, there is still much variation in the final characteristics of objects (PRINZ, ATWOOD ET AL. 1997; COOPER 2001; MAHESH, WONG ET AL. 2004; GIBSON 2005; GIBSON, ROSEN ET AL. 2010).

It can be highlighted that this technology results in porous objects, whose porosity level can vary between 50 and 90% of the density of the final object, according to process parameters and particulate material properties (PRINZ, ATWOOD ET AL. 1997 ; COOPER 2001; MAHESH, WONG ET AL. 2004; GIBSON 2005; GIBSON, ROSEN ET AL. 2010).

Figure 35 - Example of 3D printed metallic object before and after finishing

Source: Additive3D, 2012

An example of an SLS fabricated metal object can be seen in Figure 35, as well as the result of the object after the finishing and polishing process.

It is important to note that object fabricated by this type of process will have surface roughness relative to particles size. Therefore, finishing processes will be driven by material removal and coating.

Figure 36 - Example of support material for 3D metallic object

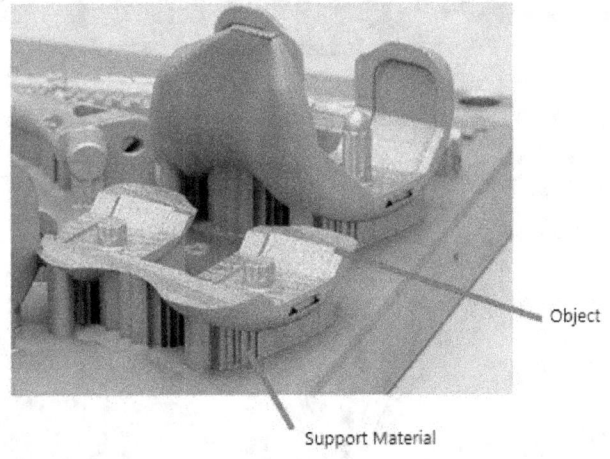

Source: LANGNAU, 2012

It can also be noted that although this process does not require support material in most cases, various metal applications indicate that the use of support material implies better object accuracy. An example of this type of application can be seen in Figure 36, where a metal object is fabricated with the aid of support material.

Just as this 3D printer technology provides a wide variety of materials, the mechanical properties resulting from this process also vary, resulting in mechanical tensile strength (yield strength) values between 5.5 and 90 MPa for plastics. (EOS 2011).

Figure 37 - Generic Schematic of Engineering LENS Manufacturing

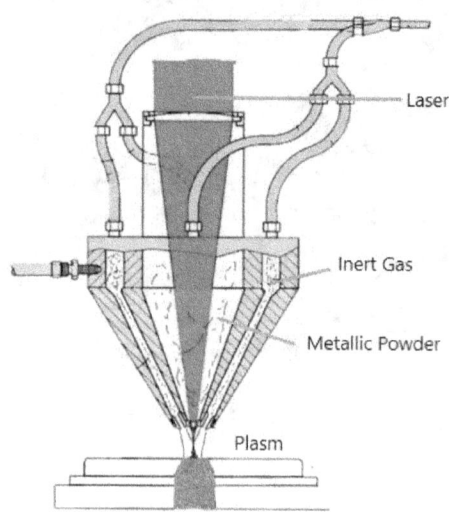

Another approach to produce metallic objects using powder and laser was firstly called Engineering LENS (Figure

37). This technology was patented by JEANTETTE in 1996 and consist in a nozzle that feed pressured metal powder while a laser beam melts powder grains. This melted powder grains become metallic liquid drops that is deposited onto surface.

It is interesting to see that this technology was almost in stand-by for years. And after patent expiration, it becomes one of the most popular metal 3D printers in the market.

Nowadays, this type of technology is the most used for hybrid equipments that combine metallic 3D printing (additive manufacturing) and subtractive manufacturing (machining).

2.4.2   3D Printing (3DP) - Zprinter

Among the various categories of 3D printers, the most widespread and that has projected the name of the technology is particle-based 3D printing. This process was initially developed by the Massachusetts University of Technology (MIT) and patented in 1989 by Emanuel Sachs and his fellow researchers.

However, the commercial product was only in the market in the 1990s by ZCorporation, which was late acquired by 3D Systems (SACHS, HAGGERTY ET AL. 1989; GIBSON, ROSEN ET AL. 2010).

The functional principle is based on the deposition of a binder under a layer of ceramic powder (usually gypsum plaster), generating an agglomerate. In this process, shown in Figure 38, a powder containing reservoir lifts a platform while a roller distributes this powder over the workpiece building platform.

For layer generation, an inkjet head moves x-y to deposit or spray the adhesive material onto the powder layer. This process is repeated until the work is finished, when an air blast is normally used to remove excess powder from the object.

Regarding the deposition thickness, this technology provides values between 0.089 to 0.2 mm, while the resolution is around 600x540 DPI. Additionally, it can be said that the overall accuracy is approximately 0.125mm (GIBSON, ROSEN ET AL. 2010).

Figure 38 - Generic illustration of zprint (3D printer based on bind-powder)

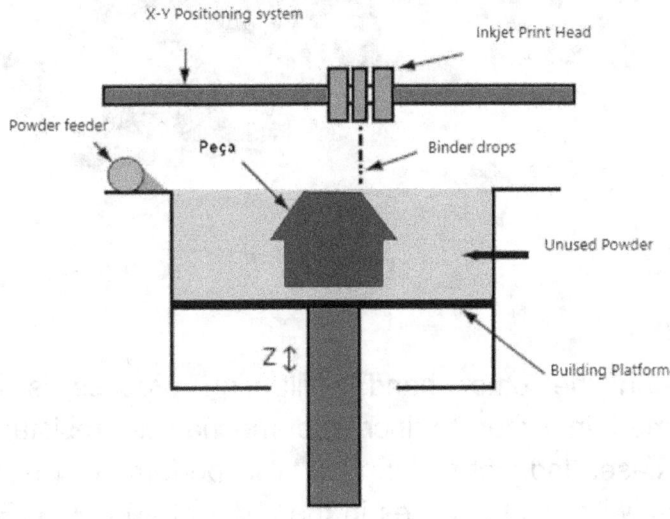

Additionally, the manufacturing speed reach 4 layers per minute, standing out from other 3D printers for fabricating color objects up to 24 color bits (GIBSON, ROSEN ET AL. 2010). An example of an object made by 3D printing can be seen in Figure 39.

Figure 39 - Example of an object fabricated by full color zprinter

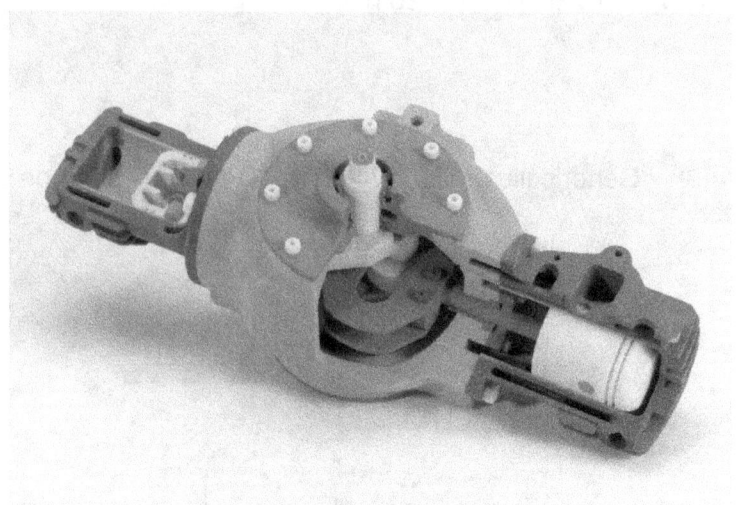

On the other hand, infiltration process is usually performed in order to increase mechanical resistance. In some cases indirect sintering is also performed, resulting in material with yield stresses in the order of 400 Mpa (EXONE 2011). However, the mechanical strength obtained by this process is low in most cases, being similar to plaster strength.

It can also be noted that, as can be seen in Figure 40, this process results in porous surface finish as a function of powder particulate material. In this figure, it is also possible to observe that the layer marks may result in uneven surface with high roughness.

Figure 40 - Example of surface condition of zprinted object

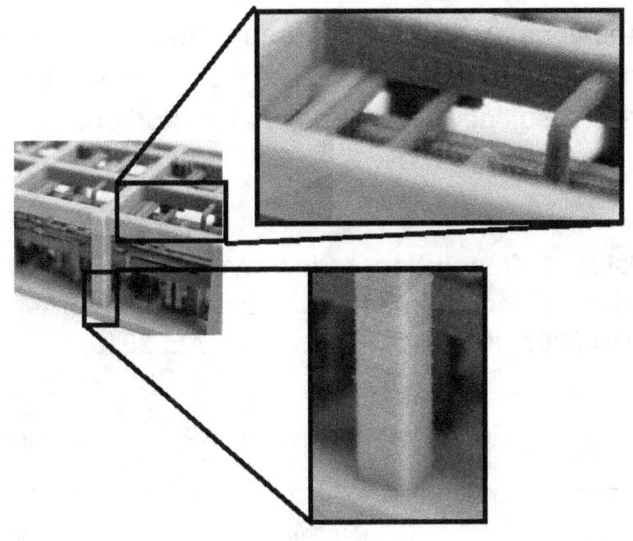

2.5 Main technnical comparison between technologies

Overall, each 3D Printer (Additive Manufacturing) process provides unique and interesting features. Thus, through the comparative analysis between these technologies, it becomes possible to identify the main benefits and disadvantages between each of these processes.

Table 1 presents a survey of the main technologies marketed in Europe, as well as their respective manufacturing resolutions and minimum layer thickness.

Additionally, it can be seen that despite the high resolution, these technologies are subject to process variations, resulting in dimensional distortions in the final product.

Table 1 Resolution ratio, layer thickness of key additive manufacturing technologies adapted from (PHAM AND GAULT 1998; LIOU 2007; KRUNIĆ, PERINIĆ ET AL. 2010)

Technology	Resolution (x-y)	Layer thickness (z)
SLA	± 100 µm	50 µm
FDM	± 127 µm	50-762 µm
LOM	± 127 µm	76-150 µm
SLS	± 51 µm	100-150 µm
3DP	± 127 µm	250 µm
Polyjet	600-1600 dpi (15- 42 µm)	16-32 µm
Envisiontec DLP	1280 x 1024 dpi (19 x 25 µm)	15-100 µm

This issues can be clearly seen in Figure 41, where the warping of a part manufactured by FDM at the end of the manufacturing process is presented (MAHESH, 2004). It is also important to identify that all technologies imply on distortion because of process parameters.

In the study by MAHESH, 2004 indicated the dimensional differences provided by the 4 most widespread AM processes today.

Figure 41 - Warpening caused by lack of adherence between platform and object

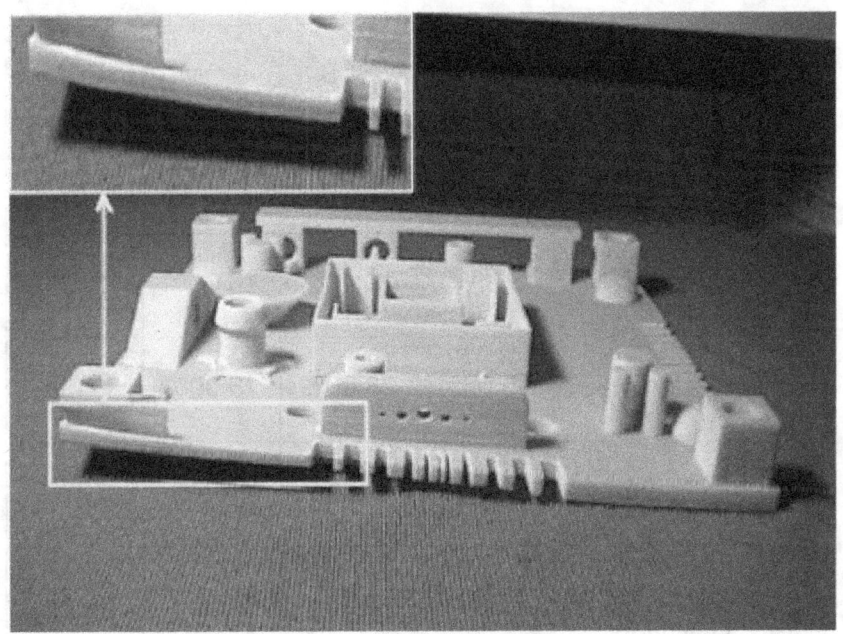

Source: MAHESH, WONG ET AL., 2004

In order to summarize the main information contained in this study, the main statistical results of this study were compile in Table 2. In this analysis stereolithography is indicated as the most accurate process among the studied.

Table 2 Statistical survey of dimensional divergences of parts manufactured by the 4 main additive processes, based on (MAHESH, WONG ET AL. 2004; CUNICO 2013)

Divergence	Additive manufacturing			
	SLA	SLS	LOM	FDM
Mean	1,33%	4,39%	5,10%	8,88%
Median	5,00%	15,00%	10,00%	12,50%
Maximum	15,00%	25,00%	25,00%	50,00%
Minimum	5,00%	5,00%	5,00%	5,00%
Standard Deviation	0,93%	3,93%	5,59%	19,93%

3 The rise of Industry 4.0

In recent years, additive manufacturing technologies, also known as 3D printers, have gained a fundamental and disruptive role in the daily lives of industries and even the personal lives of many (GIBSON, ROSEN ET AL. 2010; CUNICO 2012; CUNICO 2015; VOLPATO 2017).

It can be identified that the change in production profile and market segments has also been undergoing major changes, as demand and product customization has changed (RICHARDSON AND HAYLOCK 2012; SAURAMO 2014).

As consequence of that several movements accelerated development of industry 4.0 technologies, such as open sources communities, Do-it-yourself, Hacking, Reprap, Fab@home among other movements.

Another point that is also important to highlight is that the birth of industry 4.0 differs from other industrial revolutions because of several benefits of 3D printing, as seen in Figure 42.

It can be observed that the advancement of technology has always driven industry development, ranging from steam engines to artificial intelligence systems.

In this regard, we can identify that the first industrial revolution was marked by the mechanization of loom, creating the concept of mechanized production through machines powered by water or steam.

On the other hand, the second industrial revolution is marked by the development of series production lines, with the distribution of resources and the use of electricity.

In the third industrial revolution, the introduction to information technology systems and electronic technologies implied the automation of processes and systems.

In the fourth industrial revolution, the main core of revolution is the optimization of the productive ways through cyber-physical systems.

Figure 42 - Schematic of evolution of industrial revolutions

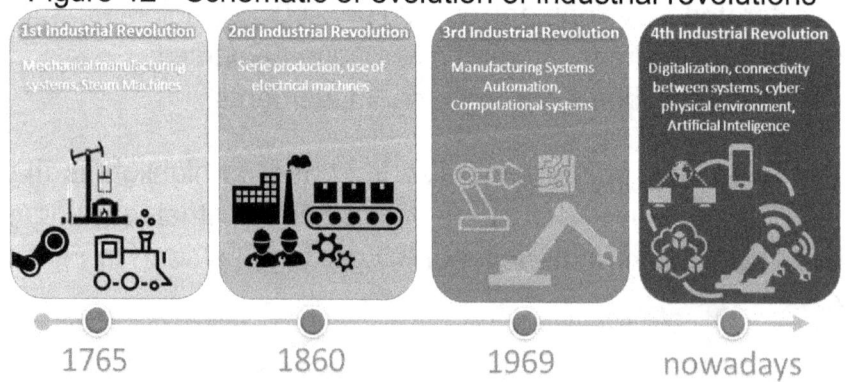

With the goal to create physical-cyber environments into Industrial environment, spheres of cultural modification and corporate infrastructure were identified: 1. Digitalization and integration to vertical and horizontal value chains; 2. Digitization of products and services offered; 3. Business model digitization and customer access, as can be seen in Figure 43.

To support these changes, several technologies have been used to systematically accelerate and develop corporate change: a) mobile devices; b) IoT platforms; c) cloud computing; d) augmented reality; e) positioning detection technologies, f) big data analysis and advanced algorithms; g) cyber security; h) smart sensors; i) autonomous robots; j) 3D printers.

Regarding the contribution of 3D printers to Industry 4.0, they are used in conjunction with 3D scanning technologies, standalone robots and rapid prototyping of electronic circuits as development accelerators, as well as productive resources in early commercialization (Startup) and at the beginning of commercial scaling (scaleup).

On the other hand, 3D printers also assist in product system maintenance, and product portfolio, whereas equipment maintenance parts can be manufactured by 3D printing technologies.

Another aspect that also underlines the use of 3D printers and open-source projects in industry 4.0 is the option of optimized replicability and infrastructure scalability. In this case, optimized replicability can be characterized as the ability of one equipment or system to manufacture another of itself with greater accuracy or capacity than the original equipment.

Figure 43 - Schematic of key industry support technologies 4.0

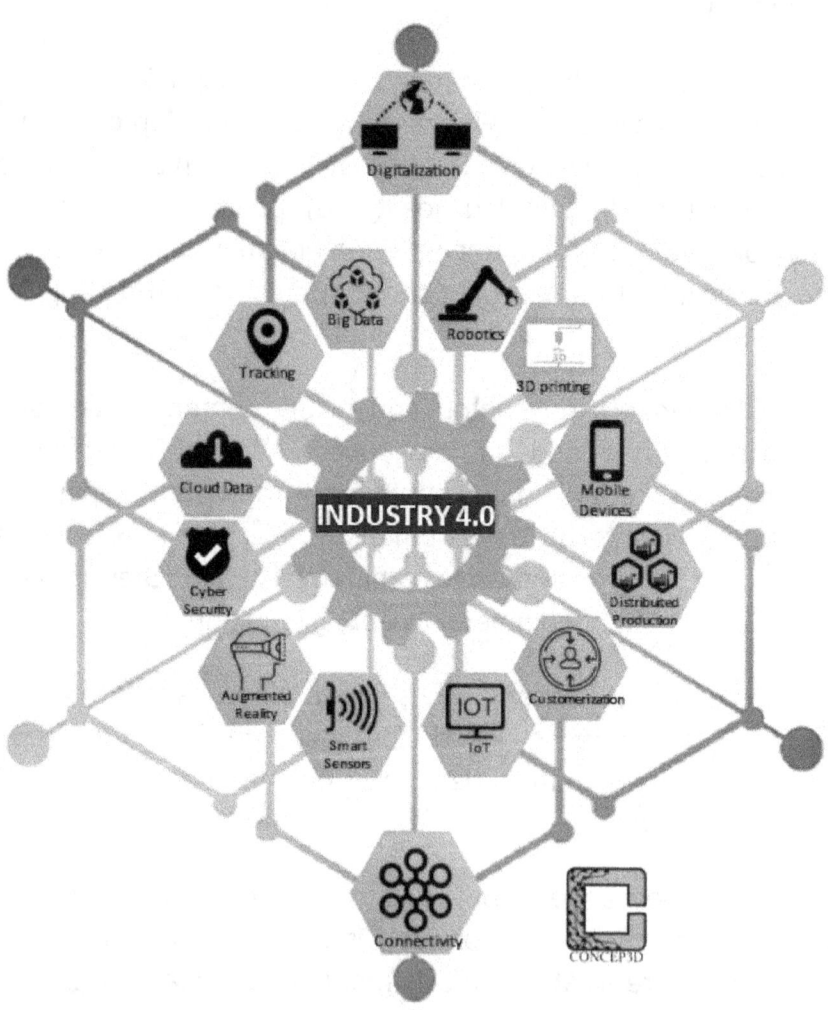

In this case, we also have the organic occurrence of distributed manufacturing centers, where there is serial manufacturing of 3D printing equipment, and offering of 3D printing services in a serial manner. Thus, it presents a very high operational efficiency rate, as can be observed Figure 44.

Figure 44 - Example manufacturing center (3D printing Farm) with over 300 3D printers placed in a 50 m² room

Source: (PRŮŠA 2018)

Note that 100% of the printers manufactured in this manufacturing cell were produced from another 3D printer.

Similarly, infrastructure scalability encourages the ability of equipment to produce other equipment without or with little need for secondary processes.

An example of this type of approach is for the manufacture of complex equipment such as micropipettes, spectrophotometers, laboratory instruments (PEARCE 2013) and also robots, as shown in Figure 45.

Figure 45 - Example of manufacturing 5 degree of freedom robotic arms through 3D printer cell

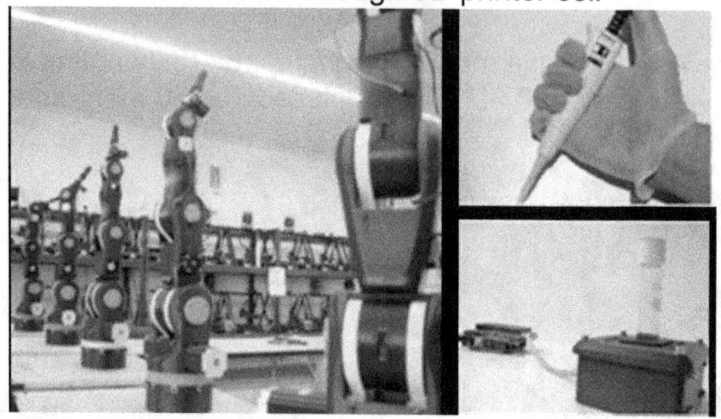

Source: Adapted from (PEARCE 2013; BCN3D 2016)

3.1 Breaking product development paradigms

It can also be pointed out that the technological movement around the development of 3d printers has also implied on changes in product concept mindset and product design development.

Once the cost of prototyping aesthetic objects has been drastically reduced, the fundamentally waterfall-like product development process can be reviewed, and agile methodologies such as extreme programming (XP) scrum, scrum-kanban (widely used in software development) become part of the daily life of designers, designers and product development engineers.

Another important point in the advent of 3D printers is the ease of making freeforms. At this point, several authors mistakenly indicate that 3D printers allow any shape to be manufactured. Unfortunately, this statement is not true.

3D printers allow you to make shapes that are by far freer than conventional processes. However, each of the 3D printing technologies has its own restrictions and characteristics so that manufacturing can be more complex and reliable.

An example of manufacturing freeform products can be seen in Figure 46, where Professor Olaf Diegel of Lund University prints musical instruments with internal free and functional mechanisms.

Figure 46 - Example of Manufacturing Guitar with Free Engines Printed 100% on SLS Equipment

Source: http://www.oddguitars.com/

Although the result is extraordinary, the process for obtaining it is very difficult, where there is still the need for removal of metallic support materials, finishing process, removal of steps and porosities through primer, paints and polishes. In many cases, even the object breaks, due to the geometric variation provided by the freedom of shapes.

For this reason, several research groups around the world are looking for design recommendations for additive manufacturing. An example of a recommendation is presented in Figure 47, where the change of design implies a reduction in manufacturing time, as well as eliminating the need to use support material.

Figure 47 - Example Design Recommendation for Support Material Minimization

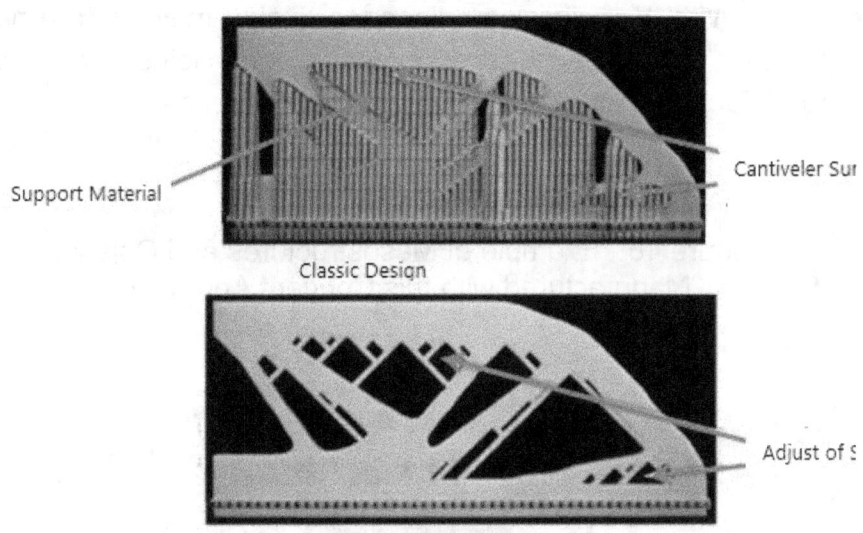

Source: Adapted from (LEARY, MERLI ET AL. 2014; THOMPSON, MORONI ET AL. 2016)

Other interesting aspects of additive manufacturing are projects based on mesostructures (lattices or Scaffolds). These design shapes also gain visibility due to 3D printers. In these design approaches, objects are manufactured with very low mass, as can be seen from the Figure 48.

It can also be noted that in this design approach, generative design is also gaining ground, where geometry is simply a result of numerical calculations for strength increase and mass reduction concomitantly.

With this, classic engineering design methods are put to the test, having to reinvent and adjust to new manufacturing, design, design and market trends.

Figure 48 - Example of Mesostructures and Objects Manufactured with this Concept Approach

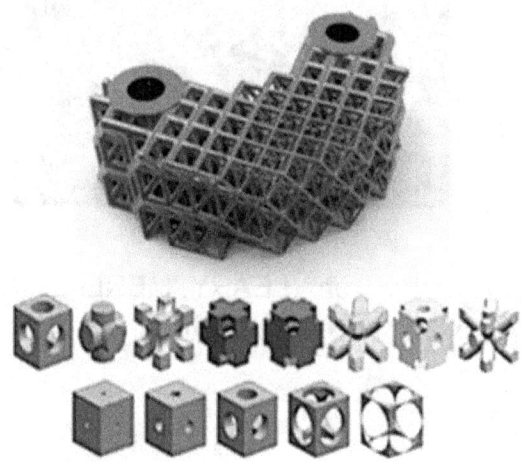

Example of Lattices, Scaffolds and mesoStructures

Source: Adapted from (LEARY, MERLI ET AL. 2014; THOMPSON, MORONI ET AL. 2016; NTOPOLOGY 2018)

3.2 Collaborative development and open-source: accelerators for industry 4.0

One of the main accelerator of industry 4.0 was 3D printing popularization. The popularization of 3d printers was not easy or fast to happen. It can be observed that 3D printers have been well established in the market since the 1990s. So why did this popularization take so long to occur?

The beginning of the popularization of 3D printers had major milestones:

- Hull and Crump patent expired in 2005 and 2009 respectively
- REPRAP group was founded in 2005 at the University of Bath - to create self-replicating 3D printers
- Birth of the FAB @ HOME group at Cornell University in 2005 - with a proposal to create open source 3D printers
- In 2006, the development of free manufacturing planning software (Slicer) at the University of Bath (England) resulted in the Skeinforge and Reprap host software that later made up 90% of the slicing slicing code on the market today.

Another milestone for the popularization of 3D printers was the open design distribution and low cost marketing of MAKERBOT printer. Marketing by itself would not imply great prominence. However, the initiative of Bre Pettis, Adam Mayer, and Zach Smith to create an online project sharing platform has led to a disruption of classic design concepts, creating the concept of collaborative design. With this, in just four years, the Makerbot company was acquired by Stratasys in 2013 for $ 604 million, leaving a legacy for the popularization of 3D printers.

We can also indicate the advent of the concept of Fablab (manufacturing laboratories) in 2001 at the Massachusetts Institute of Technology (MIT). As a result, the DIY culture has been intensified for technological things, allowing for creative and design freedom. Through this initiative, among others, we can see an accelerated development of 3D printing technologies, 3D scanning, cloud computing and standalone robots, which are key supporting technologies for industry 4.0 implementation.

The FABLAB initiative currently has as its most popular equipment: 3D printers, robotic arms, laser cutting, 3 axis CNC, printed circuit milling machine or printed circuit corrosion station. Along the same lines, PEARCE (2013) was compiled a set of open-source designs and manufacturing techniques for the creation of bio-engineering, materials science and engineering laboratories.

At this point, several other technologies have gained great prominence, enabling the viability of industry 4.0, in the form and understanding that one has today. For example, three-dimensional image mapping techniques have evolved greatly, resulting in reduced costs for implementation in production lines.

3.3 KEY 3D PRINTING APPLICATIONS FOR INDUSTRY 4.0

For many years, additive manufacturing, also known as 3D printers and rapid prototyping, has been widely used for the sole purpose of prototype manufacturing. Among the main reasons for this approach is related to the technology maturity level, as well as the patent owners' knowledge of these technologies.

It can also be identified that objects made by 3D printers have different characteristics from objects made by other processes, creating a cultural barrier for designers, designers and engineers using such resources.

In addition, timing of 3D printer manufacturing are extremely longer in comparison with other processes, implying restrictions for large-scale application.

For example, the lead time of a part made by injection mold and the same part made by FFF 3D printing are respectively 20 seconds (injection) and 2 hours (3D printer). In this same case, the mechanical strength of the 3D printing part is 70 to 80% lower than the injected part, and it has roughness on its entire surface.

So why have 3D printers had such an advent and growth? The investment for tooling and plastic injection equipment would cost $ 150,000, while an open source 3D printer currently costs $ 3,000 (50 times smaller).

3.3.1 PROTOTYPE MANUFACTURING

As for prototype manufacturing, 3D printers are already steadily occupying this chair. However, we can still identify that with the popularization of this technology, the unprofessionalization of labor occurred. With this, we have products with questionable finishes, brittle and that do not meet the customer's needs. Figure 49 presents some examples of serious defects that occur in FFF type printers, such as: a) loss of adhesion between layers; b) bubbles and extrusion heterogeneity; c) lack of fulfillment d) wrong filament distance.

Figure 49 - Example of Defects Caused in 3D Printing Process

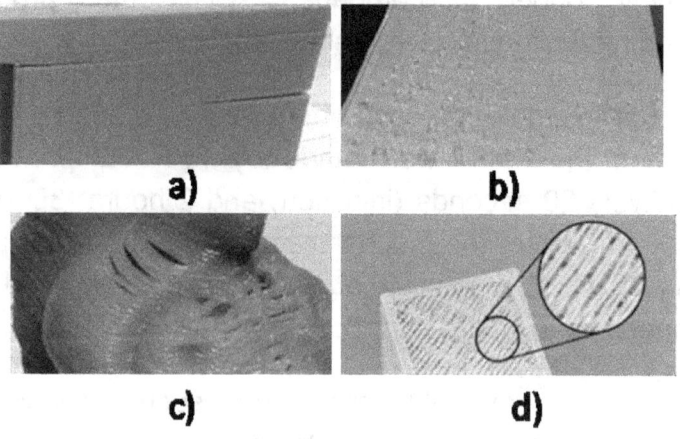

In addition to these defects, we are still facing a market where collaborative cloud designs have grown faster than product specification techniques. Therefore, when conducting a collaborative project, or using a 3D model of a cloud (such

as Thingiverse), you can send these models to 10 different manufacturers (based on 3D printers). As a result, 10 totally different parts are obtained from each other, both geometrically, finish and also in relation to mechanical strength.

For this reason, several research groups seek to standardize information related to additive manufacturing, so that it is possible to manufacture a component in a distributed manner with maximum repeatability and reliability.

3.3.2 Small and medium scale production

On the other hand, professionalized manufacturing using additive manufacturing provides several benefits, and several studies indicate the possibility of replacing conventional manufacturing process for low volume additive manufacturing.

For example, Figure 50 shows a comparison of the cost of manufacturing parts in 3 different sizes that were manufactured by injection molding, specialized additive manufacturing offices, professional additive manufacturing equipment, and an 8 low cost 3D printers in a network manufacturing cell

We can see the economical feasibility of using networked manufacturing cell 3D for 3000 parts annual demand, while offices specializing in additive manufacturing are more interesting for annual volumes of lower than 1000 parts.

Figure 50 - Example of cost amortization of 3 injection molded part sizes compared to additive manufacturing service providers, professional grade additive manufacturing equipment and production cell with 8 low cost 3D printers.

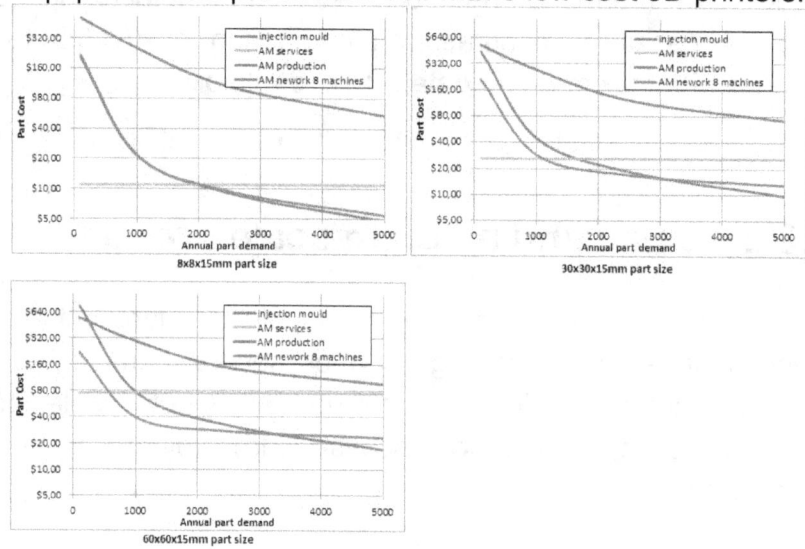

In this case, it should be noted that the total cost of the operation was compared, considering equipment, labor, rent, tooling, among other fixed and variable costs.

It can also be pointed out that in this case, only the manufacturing process of plastic parts had been analyzed, and other advantageous characteristics of additive manufacturing were not put in the agenda, such as freedom of geometric shape and object complexity.

Above all, it can be exposed that the flexibility of manufacturing complex products on demand is one of the strongest points that additive manufacturing can provide compared to conventional processes.

In this way, markets with low and medium production volume, which were previously neglected, gain competitive force. Additionally, new cyberphysical markets are created where they provide differentiated consumer experience, such as B2C marketing concepts, extremely customized manufacturing and mass customization.

For specialized markets such as aeronautics and medical, the application of additive manufacturing has become an integral part of the product development and production process. Due to the low volume, tooling and specialized production equipment can increase the cost of the product. In contrast, additive manufacturing allows manufacturing costs to remain low.

3.3.3 Hybrid Production in large scale

Other industrial application approaches are related to hybrid manufacturing processes, where the use of additive manufacturing methods and material removal methods are found in the same equipment (Figure 51).

Figure 51 - Example of CNC equipment with interchangeable

head between additive laser deposition (LMD) manufacturing and milling cutter.

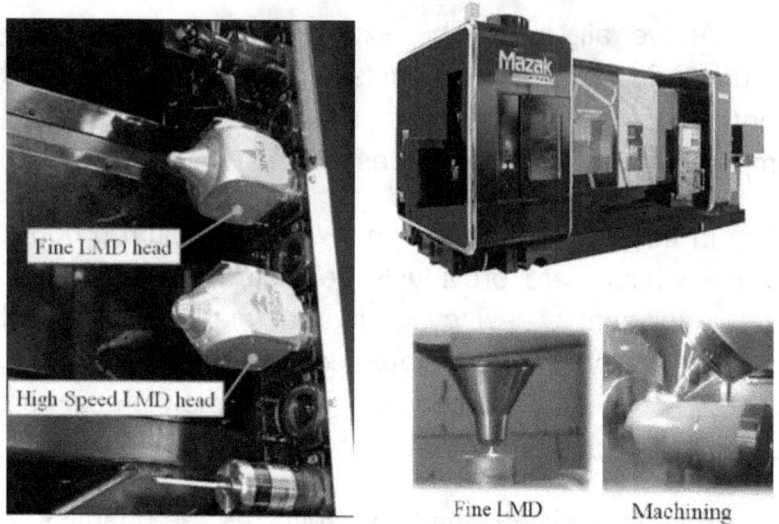

Source: Adapted from (YAMAZAKI 2016)

Other hybrid approach also considers the integration of additive manufacturing into production line. In this way, benefits from conventional processes can be utilized without giving up the flexibility of manufacturing complex objects through additive manufacturing.

Other approaches to hybrid production systems have also been studied, where tooling and support equipment manufacturing allows portfolio diversification in addition to increased productivity. Figure 52 shows a comparison of manufacturing cost by number of castings as a function of mold making process.

In this figure, we can see the feasibility of additive manufacturing over conventional large scale processes. Additionally, it is also highlighted the opportunity of portfolio diversification and increase in the number of products. As a result, increased competitiveness and proximity to latent customer needs are indicated.

Figure 52 - Comparison of Green Sand Casting Process with Wood Mold Making, Steel Mold Making, Metal Coated FDM Mold and Non Coated FDM

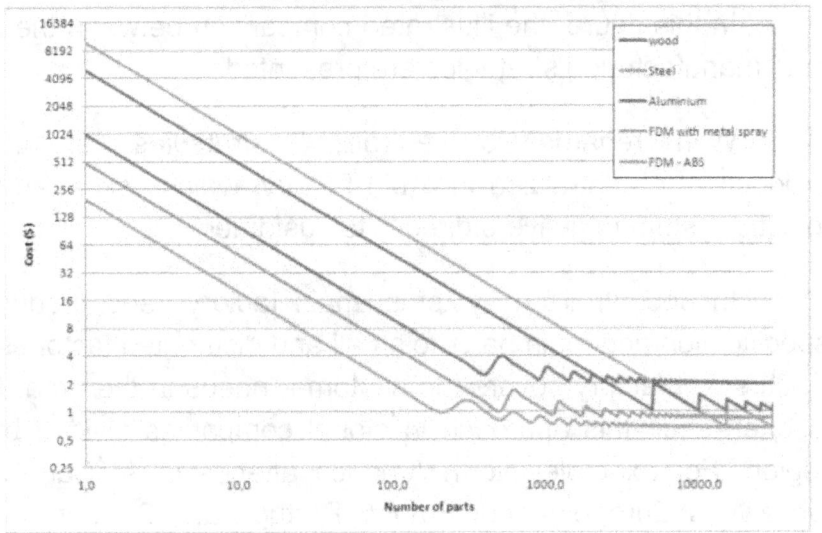

Source: (CUNICO AND KAI 2017)

In this approach, a sawtooth effect can also be observed along the cost amortization that is caused by the end of the tool lifecycle. Each sawtooth has the opportunity to completely change the product to meet new customer expectations and needs.

3.3.4 Distributed production

Another approach that additive manufacturing in collaboration with industry 4.0 presents is related to the distributed production process. In general lines, the conventional production strategy is centralized, as illustrated in Figure 53.

In this figure, the illustrated comparison between the 3 man manufacturing strategies are presented.

With regards to centralized strategies, all the production is centralized in a big factory, which use a large logistic system to delivers directly to customers.

In decentralized systems, main factory send product specification and main parts to small and distributed factories. Each small factory covers the customer needs in the area. It is often seen this approach in global companies divided by region. For example, North America attends USA, Canada while West Europe attends France, Portugal and Germany.

Although this approach have an extraordinary potential to support local market needs, several companies tend to join resources in just one factory, creating regional centralized production.

In contrast, distributed manufacturing system is brand new strategy which gain force because of 3D printing and industry 4.0. In this production model, all design, drawing and specs are in a cloud storage. Each node of manufacturing network works as logistic center, warehouse for minimal stock and manufacturing factory. In this case, it is not necessary that all nodes can produce all parts if there is a neighbor node (factory) that can supply or assembly the product.

Figure 53 - Illustration comparing centralized (a), decentralized (b) and distributed (c) manufacturing strategies

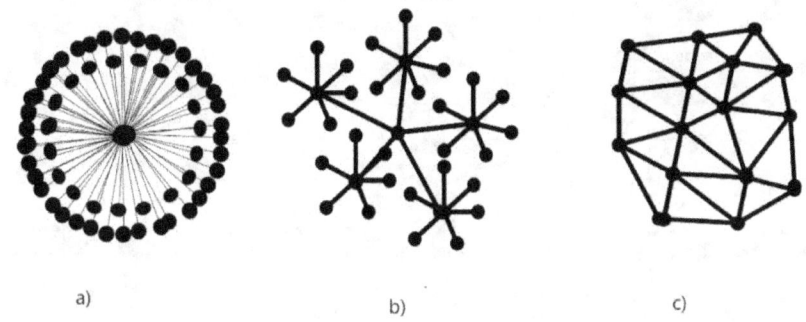

Therefore, each close loop mesh (triangle for example) is able to produce, support and deliver a family of products/parts. As a consequence, it increases flexibility, productivity and alignment of product specification with customer needs.

Unlike conventional methods, which centralize productive means to increase production volume, decentralized and distributed approaches rely on distributed FabLabs-based manufacturing centers. Thus, logistics, warehousing, inventory and infrastructure investment costs are reduced so that customers are more closely involved in development and manufacturing.

It can be seen from Figure 54 that laser cutting, robotics, and additive manufacturing technologies are currently the main options for customer-driven companies, as these technologies enable high customization and help speed product development, rapid generation of mockups, prototypes and low and medium scale production.

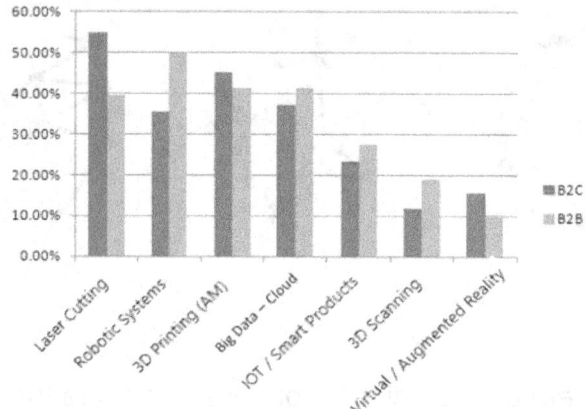

Figure 54 - Analysis of key technologies adopted by companies in B2B and B2C markets

Source: Data compiled from (BETTIOL, CAPESTRO ET AL. 2017)

3.3.5 Risks and Liability

While there are several benefits to applying distributed developments and open developments. There are currently several cases of legal issues related to the subject. Such as technical product liability, repeatability of manufacture, specification techniques.

A classic example can be observed in products with legal restrictions, such as weapons. This theme was more widespread due to the media's release of this theme in mid-2013, where a simulacrum which provides only one shot was manufactured - "Liberator handgun" by Defense Distributed (DEFDIST 2013; ROBERTS 2013), causing controversy around the world (MORELLE 2013; ROBERTS 2013; ROMANI 2013).

Other aspects of the legality of distributed projects are related to legal liability for products. For example, there is still controversy over liability when any product causes some kind of problem, injury or discomfort to a consumer or user of cloud design platforms. In conventional cases, the responsibility lies with the manufacturer and dealer / distributor. However, in this new scenario, the responsibility will be on which of the designers (when it is collaborative), or on the manufacturer, or on the raw material manufacturer, or even on the user.

These issues are still open and are being discussed extensively to identify intermediate points that do not hinder the advancement of revolutionary technology and product liability issues.

4 Applications

Due to the unique benefits and features of 3D printers, several application areas have shown interest in handling these technologies. These include: Design and Architecture; Cheers; Projects and Engineering.

We can also note that for each of these areas implies on a specific type of 3D printer as the most suitable. Therefore, it cannot be said that a 3D printer can handle all applications.

4.1 **Design and Architecture**

The areas of Design and Architecture are usually linked to the conceptual and cognitive stages of project development. As a result, a great deal of interest is observed in the complexity of shapes besides high level of details and finishing.

An example of application forms in these areas can be seen in Figure 55, where it is presented: a) building model; b) character sculpture; manufactured by a 3D printer from the 3D Systems company z-print product line.

Figure 55 - Building model (a) and sculpture (b) models made by a 3D printer from 3D Systems' z-corps product line

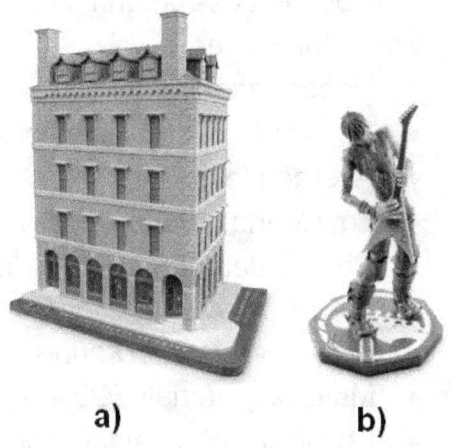

a) b)

In contrast, due to the very early stage of the project, several parameters are not considered, such as strength of materials, and requesting efforts. Therefore, there is not a very high demand for objects with respect to mechanical resistance and manufacturability.

As a result, it may be indicated that some categories of 3D printers are more suitable for the application than others.

Among the most suitable printers for application in design and architecture are:
• Particulate-based 3D printer,
- plaster based material

- Possibility of obtaining colored objects
- High level of detail
- Low Manufacturing Cost
- Post processing and finishing facility

• Color LOM Printer
- Paper based material
- Possibility of obtaining colored objects
- Average level of detail
- Low Manufacturing Cost
- Risk of damage to object due to removal of waste material

• Photopolymer based 3D printer,
- Multiple materials (Objet specific cases)
- Possibility of obtaining colored objects
- High level of detail
- The Highest Cost of Manufacturing
- Post processing and medium finish difficulty

Although there are other possibilities of use, such as FDM, SLA and SLS, these have limitations that make their application less attractive. Among the main disadvantages can be highlighted:

• FDM or FFF
- Low level of finish
- Low level of complexity due to the need for supporting material
- Average Manufacturing Cost Level
- Cannot manufacture objects with more than 3 colors (specific cases)

- SLS
  - The Highest Cost of Manufacturing
  - No manufacturing of color objects possible
  - Allows construction of highly complex objects without the need for supporting material
- SLA
  - The Highest Cost of Manufacturing
  - Painting application limitation
  - Allows the construction of object with high complexity, but needs support material

4.2 Medical and Health care

One of the larger channels of dissemination of 3D printers is health care and medical applications. This occurs due to the complexity of problems and benefits offered by these technologies.

Among the most common applications in this area is the surgical planning physical models. In this application, a virtual 3D model is constructed from computed tomography images (DICOM) and consequently used to make the physical replica (bio model) used for surgical planning.

An illustration of the main steps of this process can be seen in Figure 56.

In this figure, it can be observed that 3D models are generated from medical images, such as computed tomography. Through this 3D model, reconstruction implants, prostheses or simply the replication of organs that can assist the surgical planning can be designed.

Figure 56 - - Illustration of the main steps of the process used in the construction of biomodels for surgical planning.

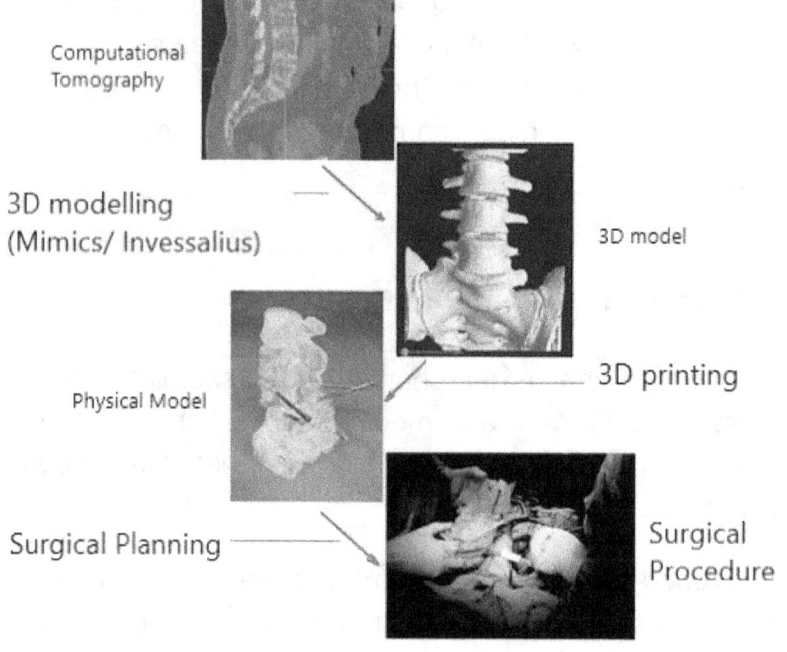

After the definition of this 3D model, additive manufacturing techniques are used directly or indirectly (molds) in order to obtain physical models that will lead to surgical planning, prostheses and implants.

It should also be noted that the use of these techniques results in less aggressive procedures to patients, in addition to reducing the occurrence of errors during surgery.

Above all, the use of additive manufacturing techniques in health care is not restricted to surgical planning models. It can also be highlighted the use of 3D printers directly or indirectly for the manufacture of implants, prostheses and medical equipment.

Figure 57 - Example of prosthesis (a) and implant (b) manufactured by SLS additive manufacturing process

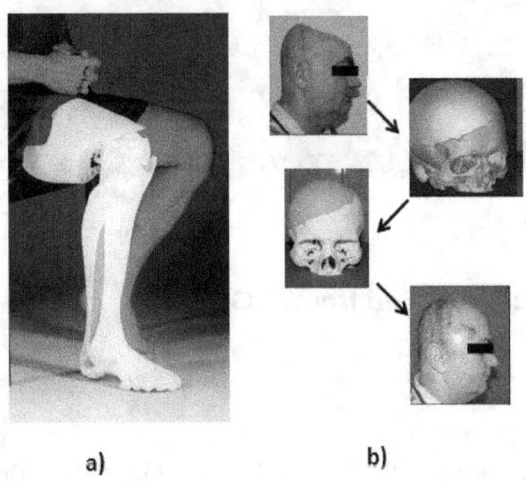

a) b)

Source: Adapted from (CARRABINE 2010; SCHUMACHER, FABBRI ET AL. 2014)

The use of 3D printing techniques is an important feature regarding prosthesis manufacturing, as the uniqueness of each prosthesis makes conventional manufacturing methods very expensive and restricted in shape.

An example of a prosthesis made of SLS can be seen in Figure 57. In this figure, an example of an implant is also shown, indicating the potential application of these technologies in medicine.

Figure 58 - Example of mandible reconstruction implant (ITG 2003)

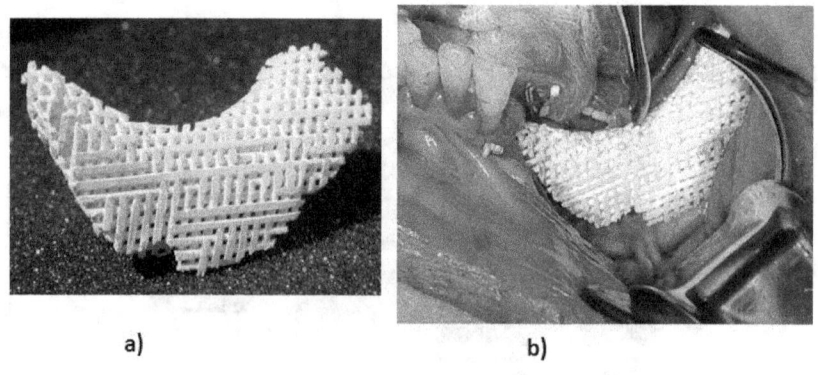

a) b)

Source: Adapted from (ITG 2003)

Additionally, it is also possible to highlight the possibility of manufacturing implants and prostheses with scaffolds, which implies on the reduction of prosthesis mass and increased implant assimilation in the body. In general lines, these scaffolds are in many cases impregnated with drugs, minimizing implant rejection and implant assimilation time in the body.

An example of an implant using scaffold techniques can be seen in Figure 58.

For the construction of biomodels, the main requirements are: high form complexity, high precision and mechanical strength.

Thus, the most suitable technologies for this application are:

• SLS

- the high level of complexity so
- High resolution and accuracy level
- Wide variety of manufacturing materials
- mechanical resistance
- The Highest Cost of Manufacturing
- No manufacturing of color objects possible
- Does not allow fabrication of enclosed hollow objects

• Photopolymer based 3D printer,

- Multiple materials (Objet specific cases)
- Possibility of obtaining colored objects
- High level of detail
- The Highest Cost of Manufacturing
- Post processing and medium finish difficulty
- Limitations on thermal distortion

- Allows manufacture of enclosed hollow objects

• SLA

- High level of resolution and accuracy
- the high level of complexity so
- The Highest Cost of Manufacturing
- Limitation of Manufacturing Material Types
- Painting application limitation
- Allows the construction of object with high complexity, but needs support material
- Does not allow fabrication of enclosed hollow objects

On the other hand, other 3D printing technologies may be used in this application. However, with more restrictions than previously mentioned:

• Particulate-based 3D printer,

- the plaster material
- Low mechanical resistance
- Possibility of obtaining colored objects
- High level of detail
- Low Manufacturing Cost
- Post processing and finishing facility
- Does not allow fabrication of enclosed hollow objects

- FDM

  - Low level of finish
  - Low level of FDM complexity due to the need for supporting material
  - High mechanical strength
  - Average Manufacturing Cost Level
  - Allows manufacture of hollow objects

- PP LOM Printer (Solid)

  - Plastic Material
  - Low detail level
  - Average Manufacturing Cost
  - Risk of damage to object due to removal of waste material
  - Does not allow fabrication of enclosed hollow objects

4.3 Engineering Design

The benefits of using additive manufacturing technologies throughout the product development process go beyond prototyping. It can also be highlighted the reduction of product launch time and design error reduction as indirect benefits of using additive manufacturing technologies.

For example, it can be identified that object handling, functional testing and assembly testing is possible to be done during the early stages of development, such as concept design. This means that the time spent on product understanding and project scope definition are reduced,. In addition, field evaluation and market analysis clinics can be performed in an accelerated.

In this case, we can make use of non-functional product prototypes, which are also known as product mockups. These mockups are presented to consumers so that a group of experts can analyze the interaction between consumers and product. Likewise, the reaction of these consumers to the product is analyzed, identifying the acceptance or rejection of the product by this public.

In relation to the conceptual design of the product, direct manipulation of physical objects tends to stimulate the creativity of developers, as well as develop product variations. Among these variations, product layout configurations that are characterized by the positioning or presence of some product feature can also be obtained faster because of additive manufacturing technologies. An example of this kind of development dynamics can be seen in Figure 59, where two product layout variations are obtained from a primary design.

With respect to detailed product design, prototypes have been noted for functionality testing, in addition to the analysis of the level of difficulty of manufacturing parts or products (product manufacturability). The use of prototypes also results in a analysis more agile than an analysis performed only in 3D software.

In this case, it can also be pointed out that the handling of prototype objects allows other stakeholders to view the product in a clear perspective. Consequently, non-technical areas such as marketing, purchasing, sales, and customer service may increase your understanding of the product in developing. Thus, input from all areas involved in the project can start in the early stages of the project.

Figure 59 - Example product layout variation from primary design

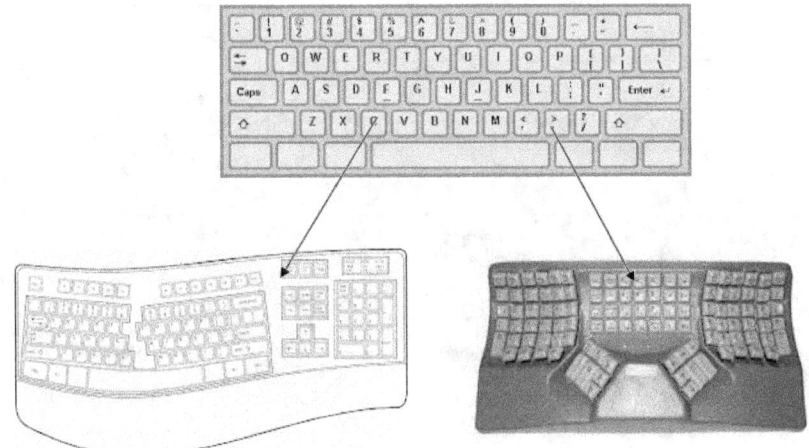

Likewise, manufacturing and assembly planning can be carried out in advanced, and the identification of product improvements in assembly and manufacturability can still be performed during detailed design, and even before component molds are manufactured.

Given this, the design time and cost related to rework and enhancements are dramatically reduced, allowing product issues to be encountered prior to release. Thus, product integrity and company identity are protected against product design errors.

Figure 60 - Comparative example of estimated average spending time to develop component design as a function of design technology

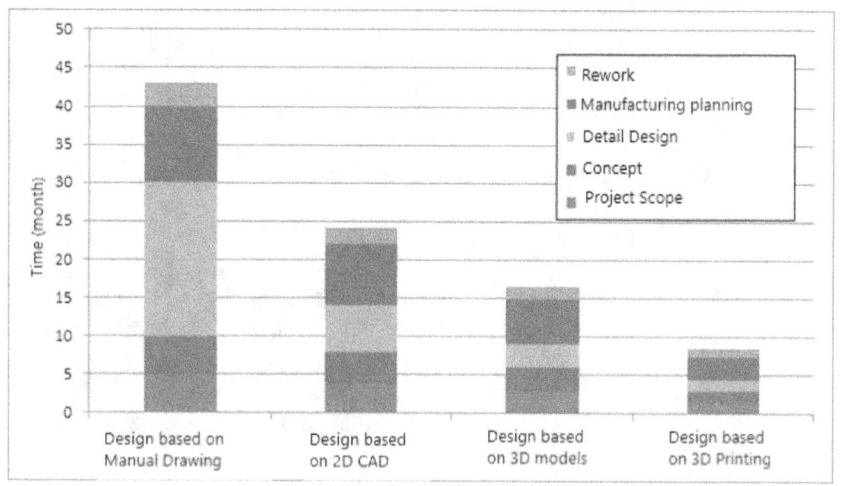

An example of reduced development time for a low complexity plastic product can be seen in Figure 60, where design techniques and total product launch time are compared.

In this figure, it can also be seen that the launch time is reduced proportionally to use of technology in design. Above all, the use of prototypes was evidenced to short product launch time.

With regards to the design and engineering areas, these are usually more related to the functional feasibility of designs and manufacturing planning. Thus, among the main characteristics required by these areas, we can highlight the mechanical resistance, dimensional tolerance and material compatibility / compatibility between prototype material and material which is used in the final process.

Therefore, it is possible to recommend FDM, SLA, SLS technologies as the most suitable 3D printers for these applications. Where we can highlight:

- SLS
    - high level of complexity
    - High resolution and accuracy level
    - Wide variety of manufacturing materials
    - Possibility of manufacturing sintered metal objects (some cases)
    - mechanical resistance
    - Dimensional error of 100μm
    - Highest Cost of Manufacturing among AM technologies
    - No manufacturing of color objects possible
    - Does not allow fabrication of enclosed hollow objects

- FDM

- Poor finishing level
- High level of repeatability
- the dimensional error of 127μm
- Limitation of Manufacturing Material Types
- Low level of complexity due to the need for supporting material
- High mechanical strength
- Medium Cost
- Allows manufacture of enclosed hollow objects

- SLA
  - High level of resolution and accuracy
  - High level of complexity
  - High Cost
  - Limitation of Manufacturing Material Types
  - Dimensional error of 100μm (AVERAGE)
  - Painting application limitation
  - Allows the construction of object with high complexity, but needs support material
  - Does not allow fabrication of enclosed hollow objects

4.4 Manufacturing and Production

Another area that also benefits from the advent of 3D printing technologies is the manufacturing and production of mechanical components such as cups, toys, small appliances, appliances, automotive parts and even aircraft parts.

Despite this, there are several groups of professionals who claim that, for the most part, manufacturing processes or production of mechanical components are not, or are barely affected by 3D printing technologies.

This statement is misleading and part of this pragmatics reflects initial classification that was assigned to additive manufacturing processes (Rapid Prototyping). Therefore, the name "prototyping" ended up stamping a feature that does not match the reality of today.

In simple terms, we can exemplify this dynamic by the comparison of the cost allocation of plastic parts which is manufactured by conventional production process (injection mold) and Additive manufacturing. In this case, investment curve and amortization by volume produced lead the specification of part price as shown in Figure 61.

In this figure you can see the comparison between the two production processes of a piece of plastic with dimensions of 150x100x30mm.

It can be identified that for low production volumes, the injection process entails an extremely high cost per part. It is caused by the high investment in mold tooling. On the other hand, increasing parts volume per year contributes to cost savings, causing part cost to reach pennies (in many cases)

Figure 61 - Comparison of part cost curve as a function of injection molding process volume and Selective Laser Sintering (SLS)

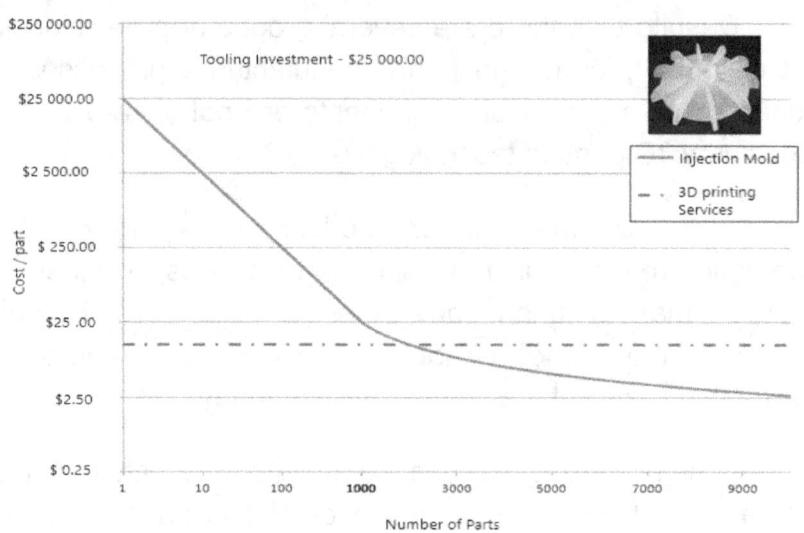

On the other hand, investment in tooling would not be necessary if manufacturing system were based on additive manufacturing (3D printing). Therefore, there would be no amortization of investment per part and the part cost as a function of production volume would be constant.

Thus, by comparing these two productive ways, it is possible to identify two distinct scenarios. For volumes less than 2020 parts, producing parts through 3D printing technologies (in this case - SLS ordered in a bureau) is the most profitable. For volumes greater than 2020 pieces, it is indicated that the injection molding process is the most profitable.

Note that these results are not the same for all part sizes or process variations. Therefore, each case must be carefully observed in order to define the process that is the most profitable to be used.

There are also two other criteria that are relevant for selecting a printing process regarding to production: the manufacturing time (takt time) and the useful area of construction of the equipment. These aspects are very important for determining lead times and batch volume.

Therefore, the most recommended 3D printing technologies for application in production pre-series or low volume production are: SLS, SLA, SLM. Note that this evaluation is related to 1 on 1 analysis.

Other processes such as FDM, IJP, MPSLA can also be considered, however with restrictions.

Among the main characteristics related to the manufacturing and production of components and products through direct 3D printers, the following stand out: finishing quality, production speed, production cost and mechanical strength.

Given this, we can recommend the additive manufacturing technologies FDM, SLA and SLS, as the most suitable for this application.

Additionally, it is noteworthy that technologies such as 3DP, LOM and IJP have restrictions related to mechanical strength, temperature resistance, resistance to weak solvents (such as alcohol) and surface finish. Therefore, the use of these technologies for the manufacture or production of components or end products is not recommended.

Among the main benefits and harms of FDM, SLA and SLS technologies for the manufacture of end components or functional parts, it can be said that:

- SLS

  - high level of complexity
  - High resolution and accuracy level
  - Wide variety of manufacturing materials
  - Possibility of manufacturing sintered metal objects (some cases)
  - mechanical resistance
  - Dimensional error of 100μm
  - The Highest Cost among Additive Manufacturing
  - No manufacturing of color objects possible
  - Does not allow fabrication of enclosed hollow objects
  - Object has surface porosity
  - Low abrasion resistance
  - Allows threading, yet susceptible to early wear
  - Low adhesion to paint
  - Does not allow the use of transparent material

- FDM

  - Low level of finish

- Possibility of finishing improvement through smoothing process
- High level of repeatability
- Dimensional error around 127μm
- Limitation of Manufacturing Material Types
- Low level of complexity due to the need for supporting material
- High mechanical strength
- Average Manufacturing Cost Level
- Allows manufacture of enclosed hollow objects
- Allow direct painting and coating
- Allows the manufacture of threads with high strength

- SLA

  - High level of resolution and accuracy
  - High level of complexity
  - High Cost
  - Limitation of Manufacturing Material Types
  - Dimensional error around 100μm
  - Painting application limitation
  - Allows the construction of object with high complexity, but needs support material
  - Does not allow fabrication of enclosed hollow objects
  - Low temperature resistance
  - Low resistance to threading
  - Enables the manufacture of transparent material

4.5 Rapid Tooling for indirect manufacturing

Another area of application where 3D printing technologies are also explored is related to soft tooling production. In this case, the final object is not fabricated by 3D printing, but by a tool or mold which is made by 3D printing. It can also be pointed out that the main manufacturing processes related to this tooling production methods are based on a copy of a pattern object.

An example of these types of production processes is Rapid Vulcanization Tooling (RVT), which can be seen in Figure 62.

In this figure, it can be seen that a mold building pattern is manufactured by 3D printing, while the final mold is used for the manufacture of pre-series parts.

In general lines, most of 3D printing technologies can be used for mold generation. However, it should be noted that the main constraints of this process are: the finishing of pattern surface and the method of manufacturing the final mold. Therefore, low cost 3D printers are not widely recommended due to the low quality and finish found on objects manufactured by this technology.

Other copy-based manufacturing processes, where 3D printers can be used, include metal casting. In these processes, metal is melted and pour into temporary molds such as sand molding, shell molding, casting or lost wax

(Investment cast) and lost plastic investment cast manufacturing. Aluminium

Figure 62 - Step illustration of temporary mold making process used for pre-series

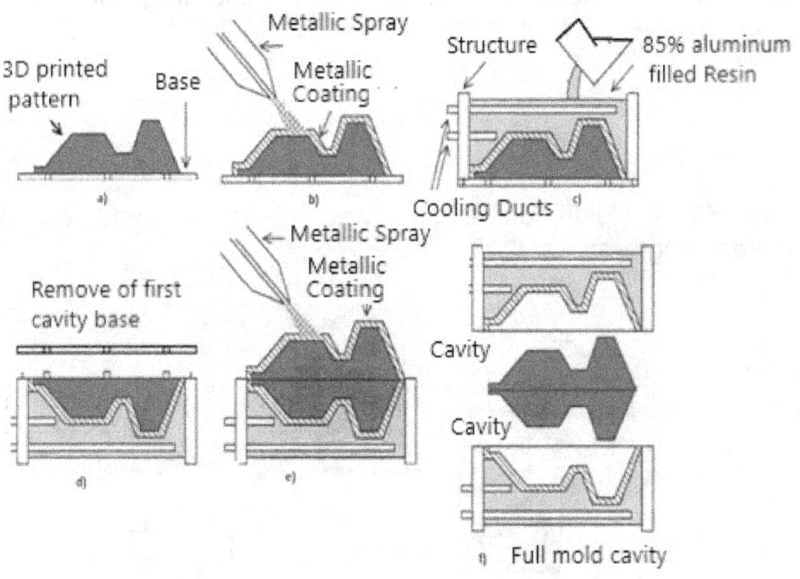

In the case of sand molds, mold patterns can be generated via 3D printers, while the rest of the manufacturing process follows the normal manufacturing flow, as can be seen in Figure 63.

In this process, the sand mold cavities are modeled through dies manufactured by 3D printers. After addition of feed and breather channels, the dies are removed and the mold is assembled. Upon completion of the mold, the metallic material may be poured into mold, filling the inner cavity of the mold.

Finally, after part cooling, the mold is broken and the sand is removed, allowing the removal of the feed and vent channels. Afterward, it is possible to finish the object through conventional processes such as grinding, blasting, machining and polishing.

Figure 63 - Illustration exemplifying steps of sand casting manufacturing process

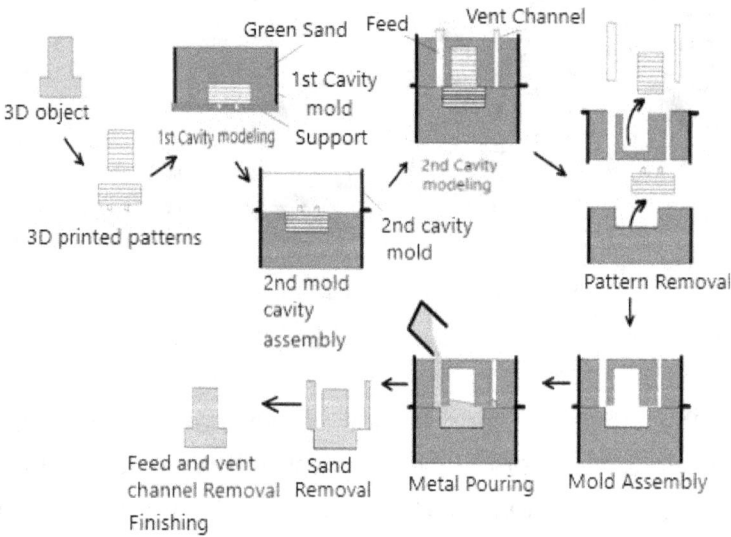

Currently, various types of parts are manufactured by sand mold, and can be exemplified:

- Hydraulic valves and fittings

- Automotive Components

- Pulleys

Figure 64 - Examples of sand casting components, as follows: a) pulley; b) automotive exhaust system; c) hydraulic connection valves; d) engine block

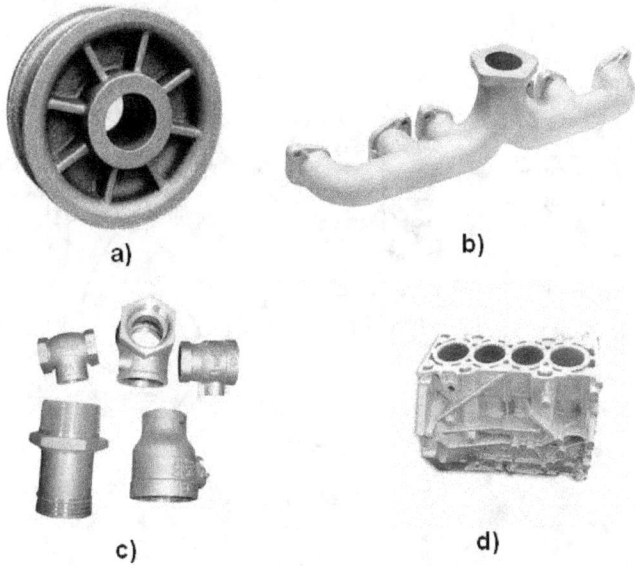

These types of components can also be seen in Figure 64, where examples of objects and components manufactured by sand molding process are presented.

Another example of a copy-based manufacturing process that can also be used in rapid tooling techniques is investment casting. This process, which is exemplified in Figure 65, is based on the manufacture of objects from shell-shaped molds.

Figure 65 - Illustration exemplifying steps of the investment casting process

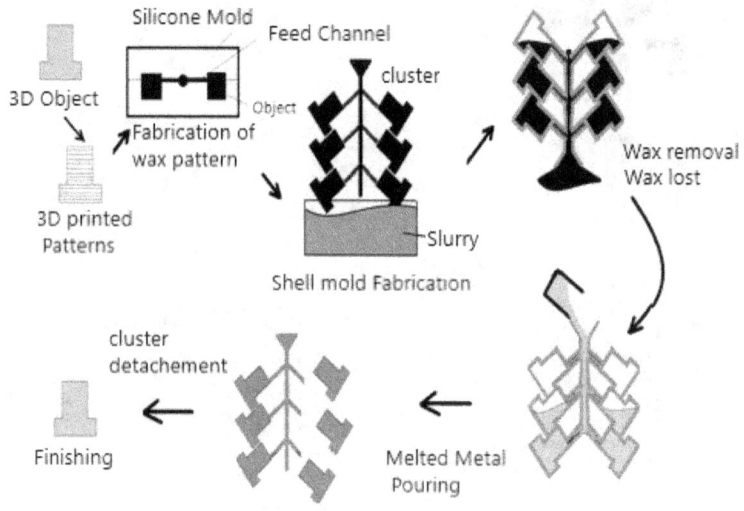

In this process, the mold is made by depositing layers of slurry (polymer solution mixture and sand) on pieces, feed channels and breaths in the pattern tree form or wax pattern, as can be seen in Figure 66.

Figure 66 - Manufacturing Cluster Example

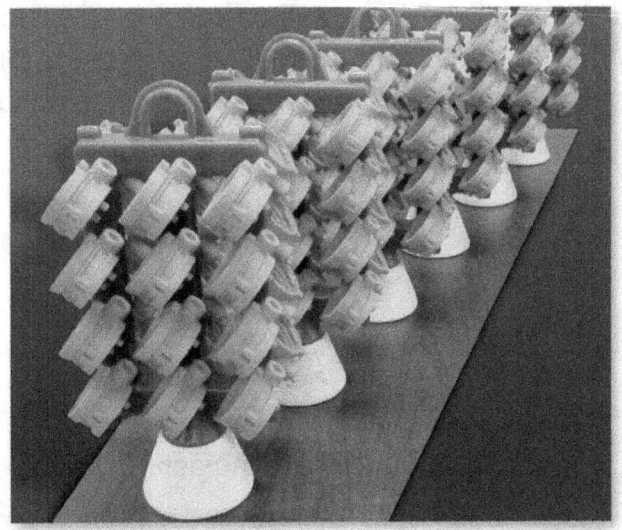

Source: (MPI 2005)

After the slurry cover is added, the pattern tree is melted and removed, leaving only the shell mold.

This way, after pouring (casing) the molten material inside the shell mold, you can break the shell, remove the parts from the pattern tree and finally finish the parts.

Regarding the fabrication of the rapid tooling, the silicone parts (dies) and molds can be manufactured by 3D printers, in some cases the building tree can be made directly by FDM or wax IJP. Thus, the fabrication of silicone mold is not necessary.

It is also important to highlight that additive manufacturing processes as well as rapid tooling techniques can also be applied to other copy-based processes that use cavities such as shell casting, vacuum casting, among others. For this reason, these processes are widely used for custom objects such as jewelry.

4.6 Rapid tooling for direct tooling

Another type of manufacturing process related to 3D printers is rapid tooling for direct manufacturing.

In this case, the final part mold is manufactured by additive manufacturing technologies such as 3D printers.

There is an incidence of using additive processes for the direct manufacture of special tools such as injection molds.

Unlike indirect tooling rapid tooling processes, this type of application is characterized by the manufacture of tools (molds) through 3D printers. This allows the mold to have geometries that are more complex than the geometries obtained by conventional processes such as machining.

The most commonly used processes for this application are based on metallic materials, due to the durability and resistance that this type of material provides. Therefore, the most used additive manufacturing technologies are: SLS, SLM, Engineering Net Shaping and Stratoconception.

Among the main benefits of this application is the customization of mold cooling channels, where the profile of the channels can follow the shape of the object.

This is a big difference compared to conventional processes as the cooling channels in these processes are straight. A comparison between these two types of cooling channels can be seen in Figure 67.

Figure 67 - Comparison between conventional (a) and custom injection mold cooling channels from additive manufacturing (b)

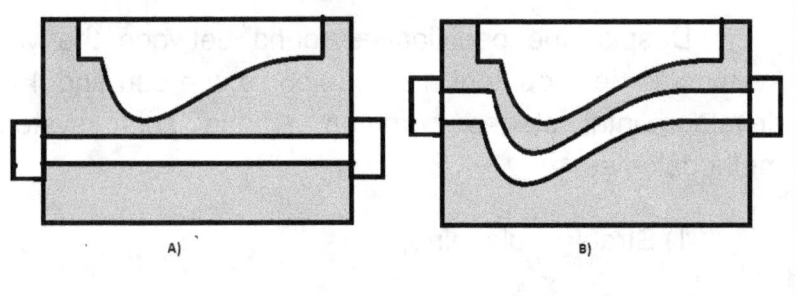

As a result, custom-shaped coolant channels provide greater thermal homogeneity in the part (reducing warpening), reduce of processing time (injection cycle), and reduce stress concentration caused by thermal difference in the mold. As result, it increases tool life and reduces production cost.

4.7 What is the most suitable 3D printer for each stage of product development?

In major lines, the product design and development process is an integral part of corporate environments. Additionally, the launch of new products plays an important role in business in order to achieve success.

Thus, several product development methodologies have been developed in order to systematize the generation of new products and reduce development time and project errors.

Despite the peculiarities found between the various systematic development methodologies, we can find 4 major developmental stages common among such systematic methodologies:

1) Strategic planning;

2) product design;

3) Project detailing;

4) Preparation for production.

In the case of non-systematic methodologies, such as design thinking, 4 main phases can be highlighted:

1) Empathy

2) Conception;

3) Evaluation

4) Preparation for production.

In this case, the various product concepts are partially or fully developed so that a design is selected to be produced.

In Figure 68, a comparative step between systematic (a) and non-systematic (b) product development methodologies is presented.

In this figure, it can be seen that systematic methodologies present a greater number of design steps and milestones, where the decision to return design steps for adjustments and improvements, as well as the project cancellation itself, can be made.

In the case of non-systematic methods, the generation, selection and refining of conceptions are performed in several rounds. In each round, the number of solutions is reduced until a proposal to be implemented for production is selected.

Regardless of the product development method, different 3D printers can be applied throughout the various stages of development. Therefore, we can correlate the needs encountered at each stage of development as a function of the benefits of each 3D printer.

Figure 68 - Comparison between stages of systematic product development and non-systematic development.

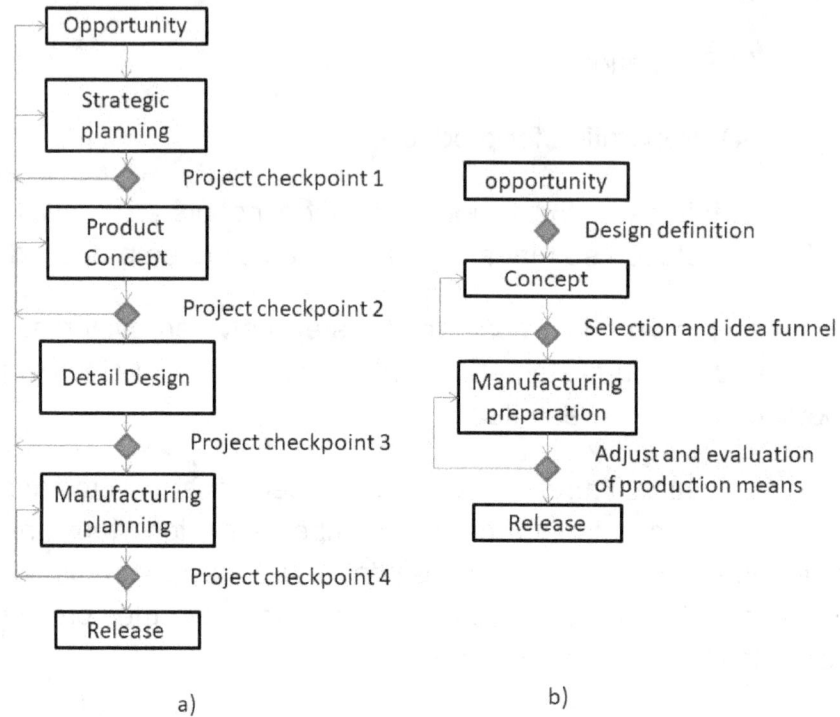

As a result, the cost spent on prototypes can be reduced and a greater number of physical models can be generated.

For non-systematic methodologies, it can be noted that this approach results in a very high development momentum, where the product design matures with each round of conception. However, this dynamism is directly dependent on the level of creativity and proactivity of the designer, design engineer or product development engineer.

For this approach, differentiation of 3D printing technologies could be attributed to the level of project maturity as well as the expected functionality.

Figure 69 - Identification of 3D printing technology according to project maturity level, type of evaluation, and prototype manufacturing cost

Design Maturity level

Shape Evaluation	Function Evaluation	Manufacturing Evaluation
3DP bind Jet***	LOM (Plastic)	FDM Industrial
LOM (Paper)*	FFF (Low Cost)***	SLS (Plastic) ***
LOM (Plastic)*	FDM Industrial ***	Stratoconception
FFF (Low Cost)	IJP	SLS(Metal) ***
FDM Industrial	SLA **	
IJP	MSLA	
SLA**	SLS (plastic) ***	
MSLA	Stratoconception	
SLS (Plastic)*	SLS(Metal)	
Stratoconception		
SLS(Metal)		

Cost ↓

* Preference

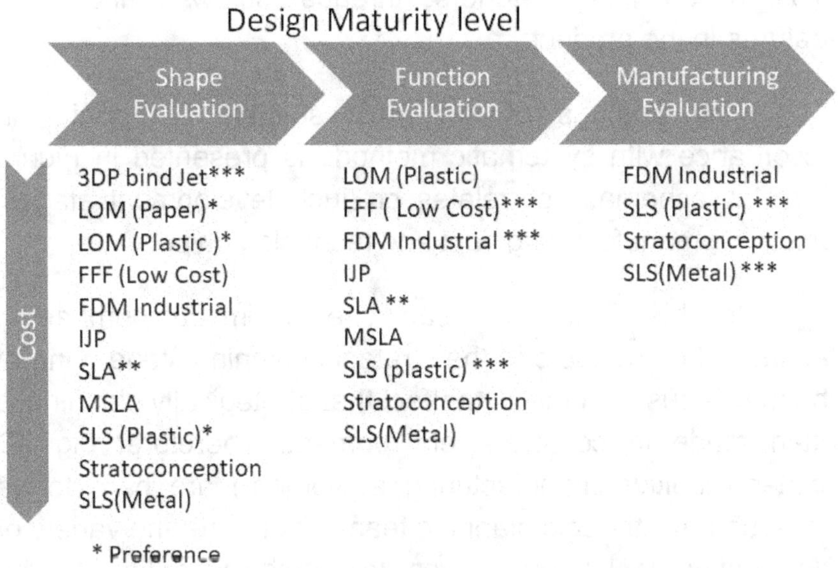

A schematic that correlates the types of 3D printing technology with design maturity level, type of functionality, and prototype manufacturing cost can be seen in Figure 69.

In this figure, it can be seen that although the number of additive manufacturing processes in the evaluation of design shape is high, the most suitable processes for this phase are particulate 3DP, SLA and LOM respectively. It is result of the manufacturing cost, level of detail that these technologies provide.

For functional product evaluation, the most recommended technologies are FDM and SLS, as these processes provide objects with good mechanical resistance.

Thus, this technologies allow designers and engineers to investigate and evaluate the functional design snap-fits, clips, assemblies, fasteners (threads, screws), and other features in the product.

The analysis of the most suitable 3D printer in accordance with systematic methods is presented in Figure 70. This schematic correlates product development stages, prototype manufacturing cost and technology type.

In this figure, we can see a limited number of technologies available in the strategic planning stage. One of the main causes of this is the fact that strategically planning is often made in corporate environment. Therefore, big 3D printer (additive manufacturing equipment) are hard to be accessed by strategic planning team. Therefore, the variety of 3D printing technologies was restricted according to the possibility of use in offices.

In contrast, the main goal of product design phase is to evaluate design shape and design functionality. In this case, 3DP and particulate SLA would be the most recommended for shape evaluation.

Figure 70 - Identification of 3D printing technology according to systematic product development step level and prototype manufacturing cost

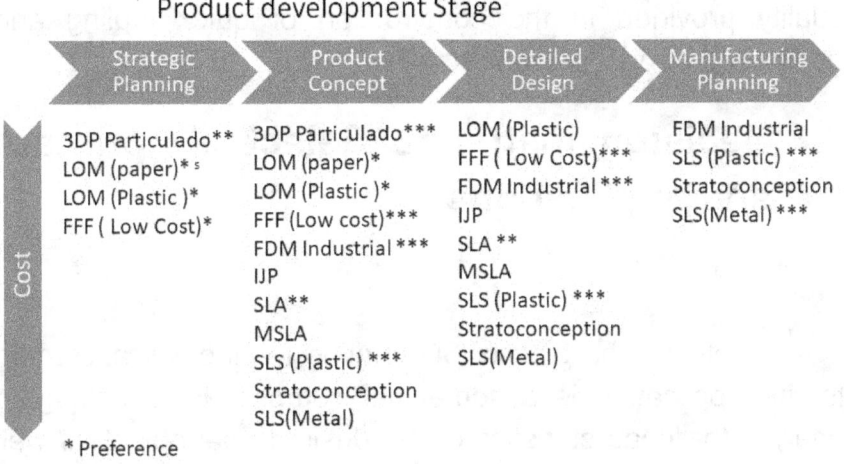

* Preference

On the other hand, SLS, SLA and FDM would be the most suitable technologies for functionality evaluation. Since these technologies manufacture objects with high mechanical resistance, making possible the fabrication of threads and assemblies.

In the project detailing stage, the main goal is related to functionality and manufacturability. For this reason, the main technologies indicated in this design stage would be SLS, SLA and FDM.

Finally, additive manufacturing technologies were restricted according to the industrial level they provide, as the preparation stage is marked by development of tooling, development of production ways and preproduction.

In this case, SLS (plastic and or metal) may be the most indicated due to the flexibility, variety of materials and quality provided in the construction of quick tooling and generation of pilot batch parts.

4.8 Decision matrix to Select 3D printers and applications

Defining the type of 3D printing technology according to the application is a somewhat complex task, having to analyze the characteristics of the desired final object as well as its expected type application.

Thus, by correlating the type of 3D printer with the final characteristic of the object, we can identify which technology is most suitable for the customer's goal.

In order to assist in the selection and selection of the most suitable 3D printing technology, a selection matrix (Figure 71) has been elaborated in order to shows main types of 3D printers as a function of the main applications and the properties that the final object is expected to have.

In this matrix, a scale of recommendations based on experiments, scientific analysis and equipment specifications compiled by the author is also presented.

In grading this scale of recommendations, the technologies were divided into 4 groups where:

• High recommendation indicates technology is the most suitable for the desired application.

• Low recommendation, indicates that the technology highly recommended for the application, but not the most appropriate.

• Neutral recommendation, indicates that the technology is not the most appropriate, yet has no severe restrictions on application.

• Not recommended, indicates that the technology in question has severe restrictions on the intended application.

Thus, the use of this decision matrix makes possible to purchase or order 3D printing services and select a vendor or technology that best suit your needs.

To use this matrix, you must first find and select the application column and desired characteristic. Then, you walk the lines of 3D printing technologies by looking at the recommendation or non-recommendation symbols. After indicating the recommendation marks, you can select the most suitable technology to use.

Figure 71 - 3D printing technology selection matrix according to application and desired characteristics

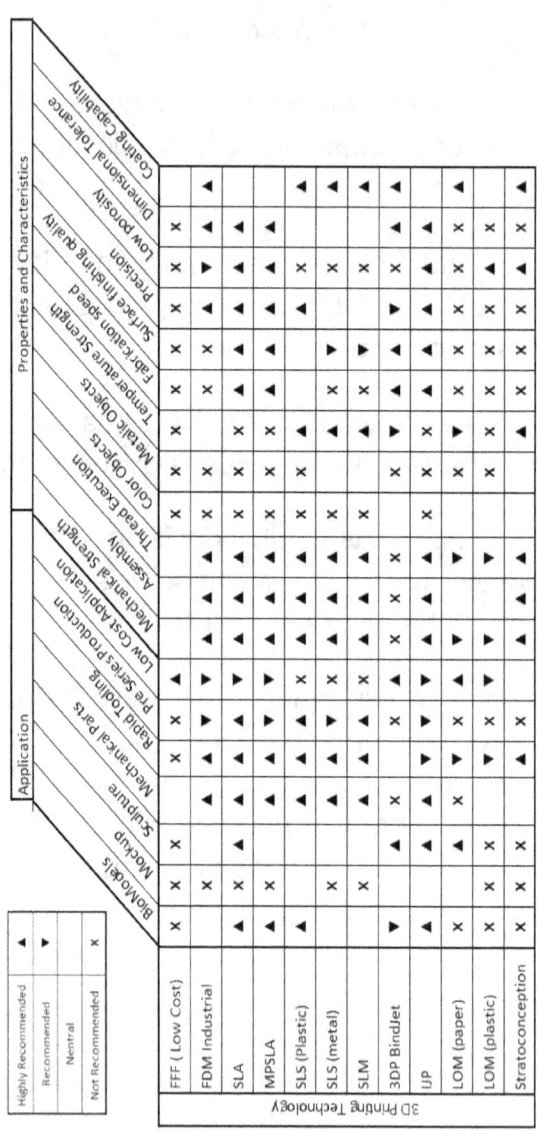

5 News, Myths and tendencies

As a closing of this issue, some of the main questions related to 3D printers are discussed, as well as some of the various researches in the global portrait and technological trends.

5.1 Myths about 3D printers

Like any new and innovative technology or process, there are many myths about usability. In the case of 3D printers, myths and even controversies related to the use of these devices have been spread globally.

Among the main myths related to 3D printers, we selected and discussed:

• 3D Printing can fabricate anything and fully Free Form Fabrication concept

• 3D printers can replace conventional manufacturing processes such as injection molding and machining

• Weapons can be fabricated at home by 3D printers

Throughout this discussion, it was also addressed and demystified the use of equipment by anyone in the home environment, indicating the main difficulties and limitations for the broadly popularization of 3D printers.

Another controversial and sensitive topic was also highlighted and discussed - the use of 3D printers to manufacture weapons in the home environment with low cost equipment. It is important to highlight that this book presents and discusses facts and data related to this subject. Therefore we kept a neutral position on the presented subject.

Likewise, this discussion is extremely important due to the fact that most of information related to the subject has several confusing sources and is often quite misleading so that this topic should be analyzed more carefully.

Throughout this chapter, we seek to present different perspectives on each of these themes, where the main objective is informative, in order to clarify doubts and potentially misconceptions generated by diffuse information of media.

5.1.1 Can 3D printer fabricate anything in any shape?

One of the main myths about 3D printers is that thes 3D printers can manufacture anything, any shape and with the best possible finish.

This myth is misleading because 3D printers only have flexibility that is greater than some conventional 3D object manufacturing processes. Therefore, it is not really possible to manufacture anything and any shape on 3D printers.

In the case of manufacturing objects using 3D printing technologies, there are many constraints, influence variables and control parameters that must be determined to make an object properly.

Figure 72 - Example of manufacturing conditions and end object result

  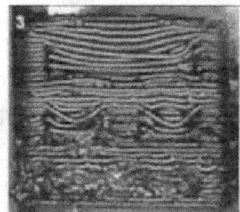

Temperature: 210°C
Extrusion Speed: 0.9mm/s
Move Speed: 16mm/s

Temperature: 210°C
Extrusion Speed: 1.1mm/s
Move Speed: 16mm/s

Temperature: 240°C
Extrusion Speed: 1.1mm/s
Move Speed: 16mm/s

Determining these parameters in a wrong way implies on obtaining distorted objects or even not obtaining them. An example of this situation can be seen in Figure 72, where a comparison between the variation of only 2 process parameters is presented.

Similarly, the freedom to construct geometric shapes depends on the type of 3D printing technology and equipment. Looking at Figure 73, it can be identified that the final result for the same object was significantly different, where several regions could not be manufactured or had a faulty finish.

Figure 73 - Comparison between parts made of (a) AFINIA H-Series; b) LulzBot AO-100

a)   b)

Source: Adapted from FRAUENFELDER, 2013

In the figure above, it can be observed that the object manufactured by process "b" presented several geometric distortions, besides a low level of finish and quality. Additionally, it is possible to indicate regions where the equipment was not able to manufacture, causing a faulty object.

Another example can be seen when analyzing FDM processes. In this case, for fabrication of cantilever geometries, it is necessary to use support material where upper layers can be supported. Otherwise the manufacture of the object does not become possible.

An example of this type of issue can be seen in Figure 74, where the need for removal of this support material upon completion of manufacture can be clearly identified.

In contrast to this issue, breakage or damage of the object during the support removal process is highly probable. Therefore, creating objects with fragile geometries and or thin thicknesses is hard to be done with support material.

Additionally, another hard of 3D printing is the fabrication of small geometries. In this case, the manufacturing size limitation is related to the deposition nozzle size, as well as the displacement acceleration and average velocity provided by the equipment.

For example, it can be identified that an configuration that is good for large objects might not be suitable for small.

Figure 74 - Example of Support Material Strategies

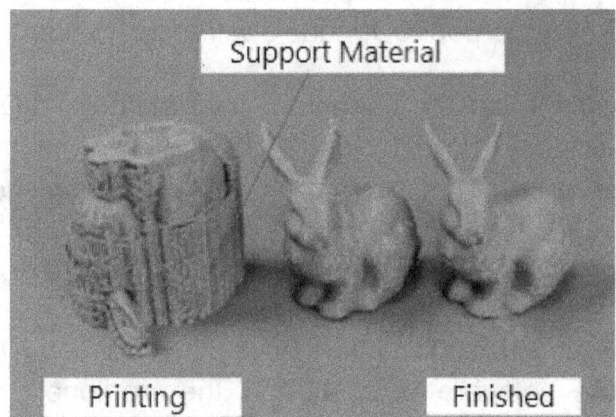

The difficulty in adjusting and calibrating the deposition profile can be seen in Figure 75, which shows the effect of extrusion rate on the deposition profile for the same layer thickness and the displacement speed of the head.

In this case, three main situations can be observed, where the high extrusion speed implies non-uniform filaments with material accumulation. At a low extrusion rate, filament formation intermittently occurs until deposition is discontinued. For combination of extrusion speed that suitable for a given thickness and head speed, the filament is homogeneous, smooth and with good adhesion.

On the other hand, it should also be considered that one of the initial stages in the manufacture of objects through 3D printing technologies is the generation of Virtual 3D models.

Consequently, there is still an intrinsic limitation for regarding to 3D modeling. Therefore, good 3D printing objects requires that users master 3D modeling systems (CAD). Otherwise poor 3D models will result in poor objects.

Thus, it can be identified that although 3D printing technologies provide a great freedom of manufacture compared to other processes, there are still restrictions. And likewise other processes, users need to be trained to avoid frustrations.

For this reason, companies like Concep3D provide proprietary software entirely in the regional language (Portuguese) and training, as well as support all consumers who purchase their products.

Figure 75 - Example of deposition profile as a function of extrusion rate variation

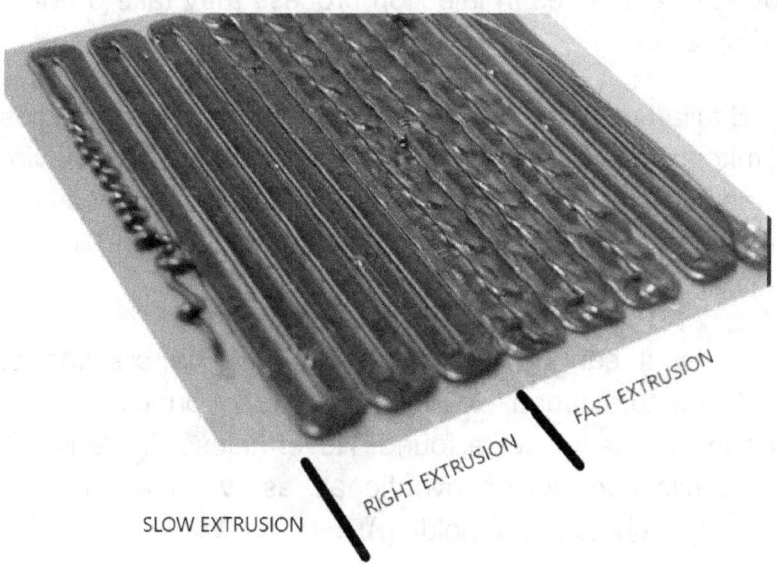

5.1.2 Will 3D printer replace conventional processes ?

Another widely spread myth about 3D printing technologies is the possibility of complete replacement of conventional processes.

In fact, despite the numerous advantages presented in additive manufacturing processes, conventional processes continue to deliver higher productivity, faster manufacturing speeds and costs that are lower than high-volume 3D printers.

For example, the manufacturing time of a part produced in 2 minutes in injection process may take 5 hours in FFF process.

Similarly, materials used in 3D printing technologies are limited and objects exhibit very anisotropic behavior (mechanical strength differs as a function of direction), causing low resistance in the direction of stacking layers (usually z).

Thus, it can be observed that with the creation of additive manufacturing technologies (3D printing), new spaces in the market were found. Nevertheless, there is still much application for conventional as well as classic processes (such as Sand Molding).

5.1.3 Can low cost 3D printers fabricate weapons ?

One of the most controversial myths related to 3D printing is the manufacture of weapons at home.

This theme was more widespread due to the media's disclosure of this theme in mid-2013.

In Figure 76 you can see a photo of the simulacrum that provides only 1 shot and caused controversy around the world (MORELLE 2013; ROBERTS 2013; ROMANI 2013).

Figure 76 - Illustration of Liberator handgun from Defense Distributed (DEFDIST 2013; ROBERTS 2013)

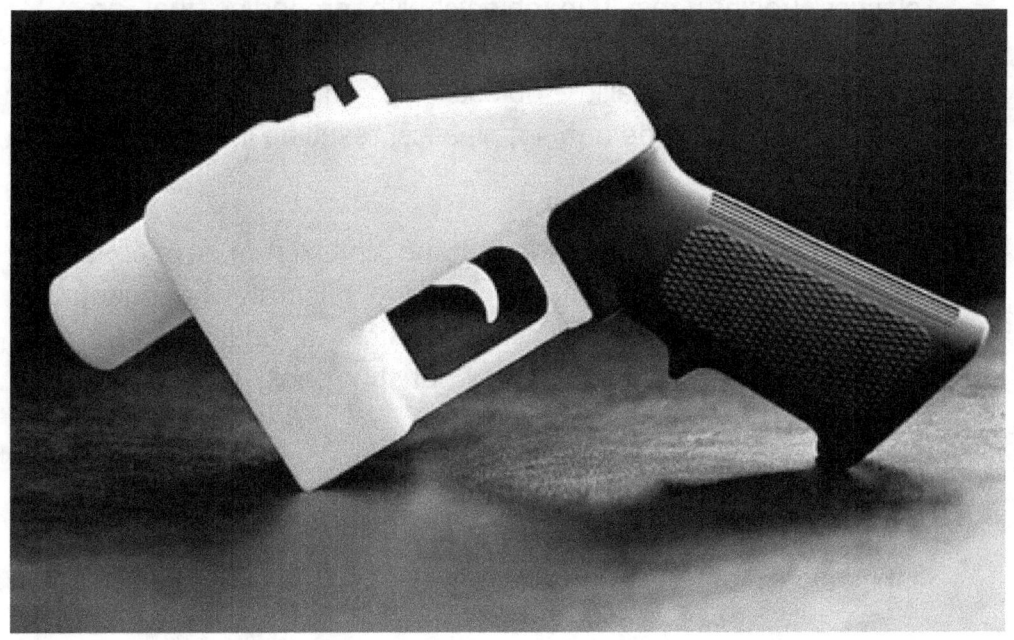

Like any other manufacturing process, there is certainly the possibility of weapons being manufactured by any of the various additive manufacturing technologies (3D printing). But despite the severity related to the popularization of weapons manufactured in the home environment, it should be emphasized that 3D printing is not the only, nor the best way to manufacture weapons in domestic environment.

As we have seen throughout this book, 3D printing technologies produce anisotropic objects (low z-resistance) and have several limitations on fabricability. In contrast, replication processes such as silicone molds, sand molds, conventional machining, and injection of thermosetting plastics (resins) result in objects with properties that are better than 3D printing objects. For example, silicone molding can produce metallic objects or high temperature thermosetting (non-deforming) plastics, resulting in very tough objects.

It should also be noted that the simulacrum developed by Cody Wilson of Defense Distributed (Liberator handgun) was not produced by low cost 3D printers. In addition to the fact that, as exemplified throughout this book, there users need to be well-instructed in order to make quality objects.

However, the focus of the problem is still pertinent, since the dissemination of 3D weapon models allows fabrication in any classic process.

Simply by making available 3D models of armaments, it is possible to manufacture it directly from metal by sand casting or even machining (HOMEGUNSMITHING 2008; BENEATHTHEFALLEN 2010; DYIGUNS 2013).

In the case of machining, we face a situation that is more aggravating, since the cost of a small CNC for hobbyists reaches $ 1,000.00 and can even be manufactured at home (WILLIAMS 2003).

Figure 77 - Illustration of homemade foundry rifle piece

Source: Adaptated from HOMEGUNSMITHING, 2008

An example of this type of application can be seen in Figure 77, where a rifle piece was made fully metal in the home environment.

On the other hand, it should be emphasized the replication processes do not even need 3D model to reproduction weapon parts directly in metal or high strength resins. However, these methods still need a model as a base.

Thus, the problem related to weapons is in a higher sphere and is not direct related to 3D printing, but rather the legalization and commercialization control of 3D models of weapons.

5.2 Researches and Innovation

With regards to researches and innovations in the area of additive manufacturing (3D Printers), it can be indicated that tremendous advance has occurred in recent years. With this, several areas could benefit and it was possible to identify the great economic potential of these technologies.

As a result, accelerated product development and short time product launching provides a competitive edge and enhances profitability of operations and business.

One of the most obvious ways to observe this increase in researches related to 3D printers and additive manufacturing is the number of patents generated over time.

Of course, this is not a direct way to measure the amount of investment in research. However, this analysis indicates the efforts of companies to innovate. Therefore, this innovation effort can be related to research investment.

Currently, it is possible to restrict patent search and monitoring by knowledge area, such as additive manufacturing (AM). These patents can be identified by company name or international patent classification (IPC), with B29C being the area of additive manufacturing and 3D printing (WIPO 2011).

By monitoring this section, shown in Figure 78, it becomes possible to identify the progression of the number of patents related to AM technologies over the last ten years.

Figure 78 - Growth in number of patents related to additive manufacturing (AM) - International Patent Classification (IPT) B29C, and of the top 5 AM companies

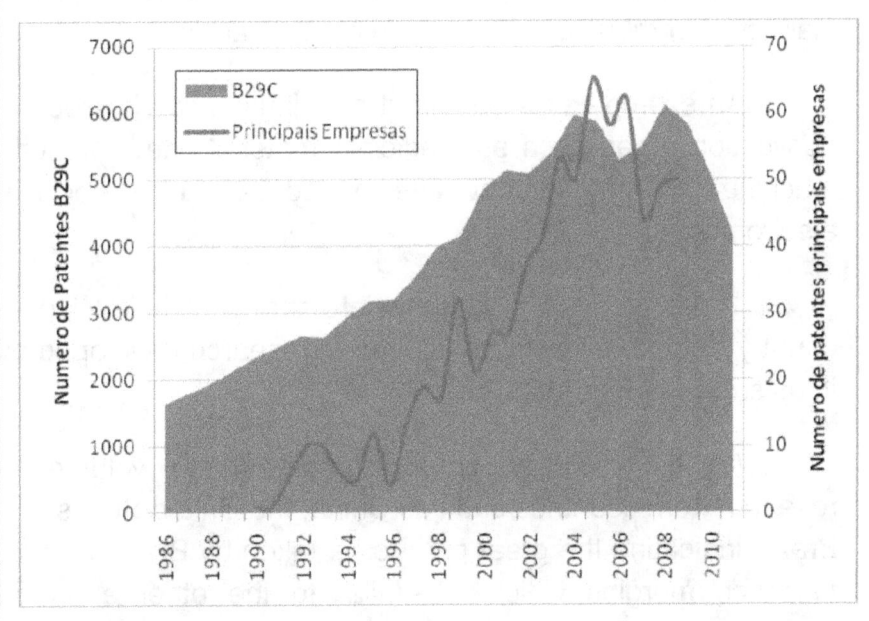

Source: Adapted from (WIPO 2011; CUNICO 2013)

This indicates that in addition to the high acceptance rate of 3D printers in the market, companies generally have very high application potential. Thus, the large number of patents generated in the last few year may also be the result of heavy investment related to additive manufacturing processes.

Additionally, this figure also shows the progress of the number of patents generated by leading AM suppliers, such as Stratasys, EOS, Envisiontec and 3DSystems.

It should also be noted that the contribution of research goes beyond the supply of a local need, this development indirectly contributes to the growth of other areas.

An example of this indirect contribution can be seen in application areas such as medicine, materials development, machine and equipment design, control and precision electronics.

Thus, we can also associate the growth level of a country's economy with the volume of research developed by it, as shown in Figure 79.

Additionally, this figure compares the volume of research from countries such as China, the United States and Brazil, indicating the great path to be taken by Brazil in order to reach margins which is similar to the other analyzed countries. Thus, it can be clearly seen the relationship between research and development efforts and the country's economic growth.

In other words, besides supplying a demand from the domestic market, the development of an entirely national AM technology contributes to the development of other areas, and consequently to the growth of the country.

Since the number of research and patents related to 3D printing technologies is tremendously large, in addition to the similarities between such a large volume of data, this book has attempted to compile some of the most relevant research and patents of the past 5 years.

Figure 79 - Relationship between research volume developed and economic growth

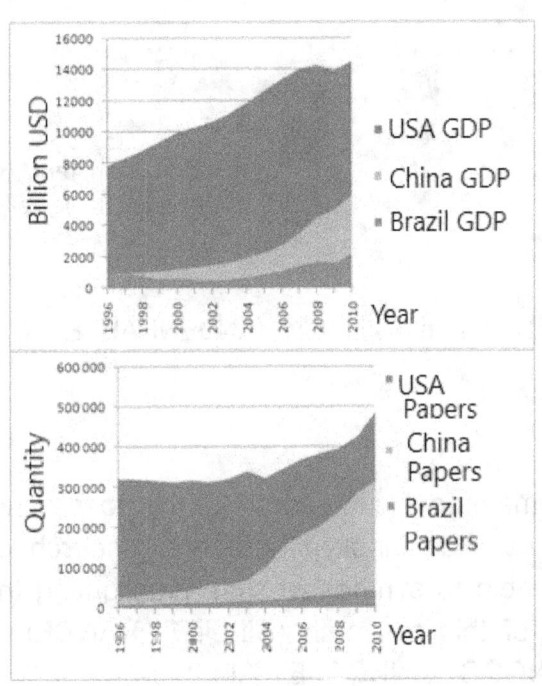

Source: Adapted from (IMF 2011; SCIMAGO 2011; CUNICO 2013)

5.2.1 BioPrinting or Tissue 3D printing

This research line has tissue generation as its main goal. In this case, a moving head deposits living cells stored in gel on substrate, as observed in Figure 80.

Figure 80 - Representation of cell and biomaterials printing system

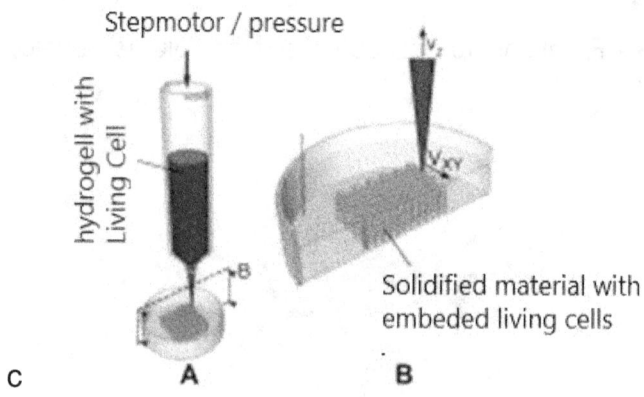

Source: Adapted from (BILLIET, VANDENHAUTE ET AL.)

The main goal is to search for the possibility of building whole organs. Additionally, there are research groups that have been able to synthesize skin, highlighting the potential application of this research (BILLIET, VANDENHAUTE ET AL.; DOMINGOS, DINUCCI ET AL. 2009).

It is important to note that babysteps are daily made in this field since 1999, when the first tissue was printed. For example, we can see that a non-function reduced scale heart was printed in order to present the advance of organize living cells in a functional shape in addition to attract funds and resources to continue the development of this wonderful area.

Figure 81 - Photo of the non-functional reduced scale heart printed with living heart and steam cells

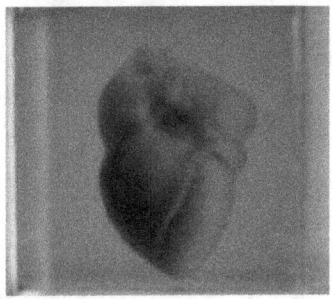

Source: NOOR, N., A. SHAPIRA, 2019

Overall, this technique aims at an extraordinarily large advance, allowing for the transplantation of vital organs, such as the heart, without the need for a donor.

Despite this futuristic ideal, research still has many challenges to overcome in order to achieve this goal.

5.2.2 SLS with multiple material

Another striking research in the area of additive manufacturing is the manufacture of multi-material and functional gradient parts.

Figure 82 - Representative multi-material laser selective sintering system and functional gradient

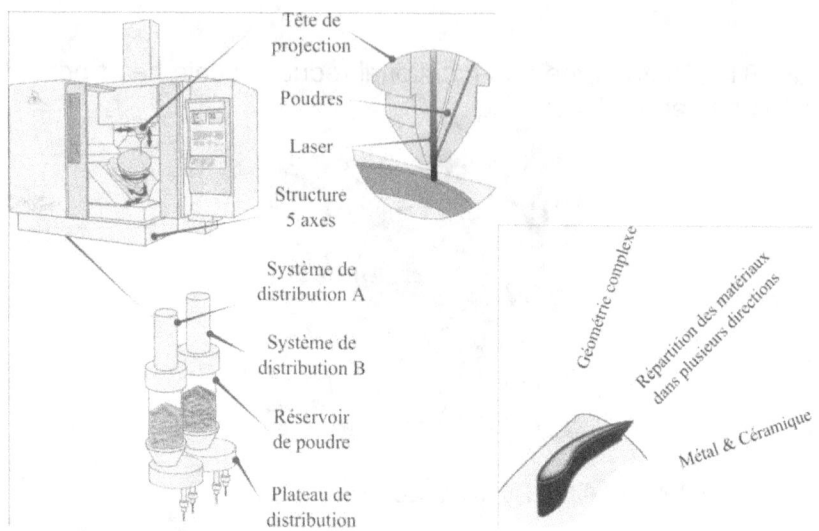

Source: Adapted from (MULLER, MOGNOL ET AL. 2012; MULLER 2013)

In this matter, the main objective is the manufacturing of objects that have different mechanical behavior, but in a controlled way.

One of the examples of this type of research is presented in Figure 82, where an additive manufacturing system (3D printing) was developed to allow the selective deposition of metallic powder from each layer in addition to different materials (MULLER, MOGNOL ET AL 2012; MULLER 2013).

After deposition of this powder layer, a laser projection head sinterizes or melts the material as x-y displacement (MULLER, MOGNOL ET AL. 2012; MULLER 2013).

Thus, it makes possible to fabricate metal, plastic and ceramic gradient parts according to the structural needs of the part (MULLER, MOGNOL ET AL. 2012; MULLER 2013).

5.2.3 3D printers in Civil Engineering

Just as 3D printing provided a great advent of new mechanical fabrication concepts, other areas of knowledge were also inspired, extrapolating scale design and application.

Among the various cases studied, the two most relevant cases are: Contour Craft and Large Scale Cement Printer (D-Shape).

In both cases, the focus is on the manufacture of very large scale objects. This concepts' vision aim to manufacture automatically even buildings.

In the case of Contour Craft, patented by Paul Bosscher and Robert Williams, a special concrete deposition system is secured by means of temporary support towers, as shown in Figure 83 (BOSSCHER AND ROBERT 2007).

Through these towers, the deposition head is moved and the construction of very large scale objects and even buildings becomes to be possible (BOSSCHER AND ROBERT 2007).

Figure 83 - Representation of Contour Crafting

Source: (COUTOURCRAFTING 2014)

This type of system is interesting because of the low material waste that is provided. However, the direct application of this type of technology still requires a lot of work, since the insertion of metal reinforcement to increase the structural strength of a building has not yet been possible to perform automatically.

Additionally, this process is still very expensive depending on the type of material deposited. Because only one type of material is still used for construction in this technology.

Figure 84 - Contour Crafting System Representation

(COUTOURCRAFTING 2014)

Source: (COUTOURCRAFTING 2014)

On the other hand, there are other variations of technology where the moving system can be carried out by columns or even portable systems as shown in Figure 85. This provides the future possibility of building infrastructure in unhealthy environments such as space. and ocean.

On the other hand, the advances in 3D printing applied in civil engineering is large and has a lot of potential, even though plenty of challenges have to e overcome yet.

In the case of Enrico Dini's patented Large Scale Cement Printer (D-Shape), a deposition system applies an aggregate material (powder composed of cement, sand, gravel among others) along one layer, while a spray selectively impregnates the substrate with binder material, as observed in Figure 85(DINI 2008).

Figure 85 - Schematic representation of D-Shape additive manufacturing system operation (DINI 2008)

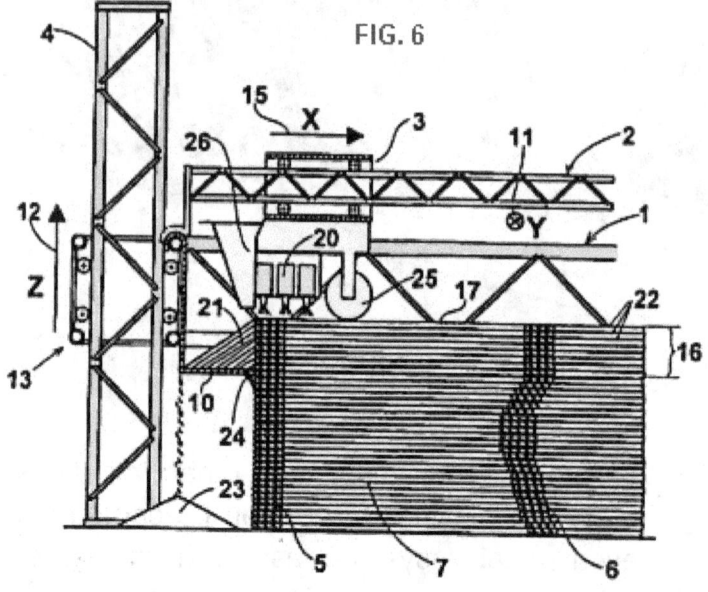

Source: Adapted from Patent (DINI 2008)

A picture of the implementation of this technology can be seen in Figure 86, where it is possible to observe the complexity related to the use of this process in a viable way.

In this way, the region where the binder material is spraid solidifies. At the end of the process, the material not impregnated with binder liquid is removed to obtain only the desired finished object.

Figure 86 - D-Shape additive manufacturing system representation

Source: (ESA 2013)

This process is very similar to particulate material 3D printing technology, however, this is performed on a gigantic scale and with differentiated materials (DINI 2008).

Despite the simplicity of the concept, the implementation of this type of technology in civil construction (houses and buildings) is difficult, since there is always a need to remove residual or non-solidified material. It is also noteworthy that building a house.

5.2.4 Simultaneous Deposition and Polymerization – SDP

Another technology that has proved to be relevant in the area of additive manufacturing (3D printing) is Simultaneous Deposition and Polymerization technology.

In this process, patented by Louis Laberge-Lebel of the University of Montreal, a liquid photopolymer deposition system is continuously deposited through a single extrusion nozzle while being solidified simultaneously at the outlet (LEBEL, AISSA ET AL.; LABERGE -LEBEL, THERRIAULT ET AL. 2008).

Material solidification occurs due to exposure to ultraviolet light as material exits the nozzle simultaneously (LEBEL, AISSA ET AL.; LABERGE-LEBEL, THERRIAULT ET AL. 2008).

Despite the similarity with photopolymerizable material (IJP) 3D printing systems, which deposit droplets of material, this process provides the generation of layer profile through filaments, just as FDM systems.

Additionally, this process is also globally researched, where materials, equipment and process optimization have been developed (CUNICO 2011).

Benefits of this process include increased object homogeneity due to greater interaction between filaments and layers and low process cost compared to laser-based technologies such as SLA (CUNICO 2011). It is also important to highlight that this technology allows the fabrication of ceramic and metallic composites.

Figure 87 - Simultaneous deposition and polymerization system representation (LEBEL, AISSA ET AL.; LABERGE-LEBEL, THERRIAULT ET AL. 2008)

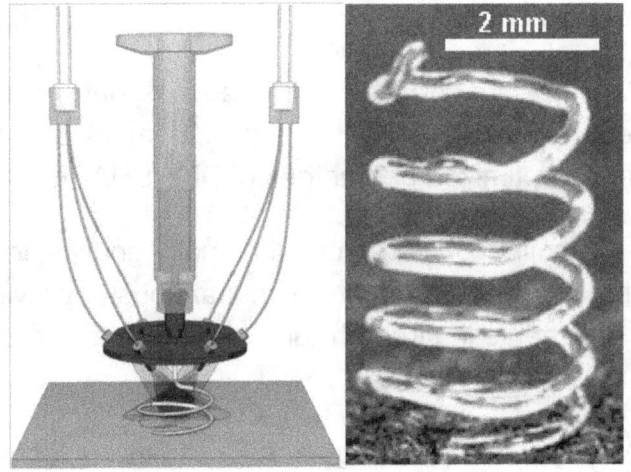

Source: (LEBEL, AISSA ET AL.; LABERGE-LEBEL, THERRIAULT ET AL. 2008)

In addition, the high bond between filaments is due to the deposited material not being 100% solidified. Therefore, a significant improvement is provided compared to FDM and IJP processes (CUNICO 2011).

5.2.5 Generative Design and Topological Optimization

Due to the significant increase in the geometry freedom afforded by the additive manufacturing process, a scientific and industrial movement has begun to revise the current design concepts.

Among the researches carried out within this movement, we can highlight the development of parts with topologically optimized geometries (POLY-SHAPE 2013).

In this line of research, the main goal is to identify geometry that, for example, simultaneously provides the highest mechanical strength and the lowest mass at the same.

As an example, Figure 88 shows a comparative representation between a topologically optimized part and one designed using the conventional method.

In this case, both parts result in equivalent mechanical strengths, and the topologically optimized part uses only 70% of the raw material required by the other (POLY-SHAPE 2013). This indicates the possibility of designing products in a new different way.

Thus, it becomes possible to design parts only indicating the general constraint, limits, interface points and couplings. After this definition, the internal geometry would suit most appropriately to provide greater strength and lower volume or weight.

Figure 88 - Comparative representation of topologically optimized geometry part and conventionally designed part

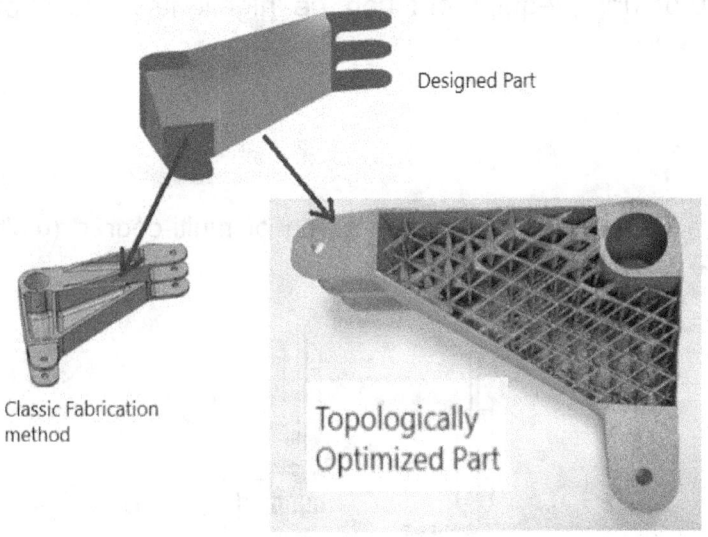

Source: Adapted from (POLY-SHAPE 2013)

It can also be observed that the way to design parts for specialized applications, such as aeronautics (which constantly looks for components with lighter weight), tends to be modified.

5.2.6 Additive manufacturing with multiple degree of freedom

Just as conventional manufacturing processes have undergone improvements, additive processes tend to change in order to provide differentiated characteristics.

In this case, the development of multi-axis additive manufacturing equipment can be highlighted, as shown in Figure 89.

Figure 89 - Simplified representation of multi-degree (multi-axis) additive manufacturing process

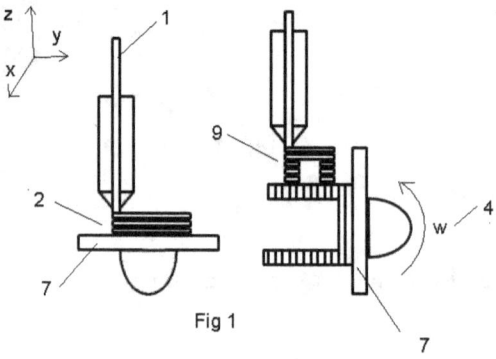

Source: Patent (CUNICO 2013)

This machine, whose pendent patent is owned by the company Concep3D, combines unique features compared to conventional 3D printing processes. Once the object manufacture does not only occur layer-on-layer only in a plane, the base surfaces of the objects do not necessarily have to be flat, and additionally, intermediate depositions can be made in order to provide different finishes.

In this process, the fabrication of objects occurs additively on more than one construction plane, and thus the shapes of the layers are allowed to be more complex, such as cylindrical (CUNICO 2013).

Thus, it is possible to manufacture cylindrical parts with greater precision and degree of complexity and co-deposition on molten or injected raw material, relief printing of objects on prefabricated surface (CUNICO 2013).

Similarly, a finishing deposition process can be considered, where three-dimensional layers are deposited so that the unpleasant ladder effect caused by the conventional layer-by-layer process is minimized (CUNICO 2013).

5.2.7 Multiple colors FFF

Other research that has also been relevant over the past 5 years has been the development of FFF 3D printers that deliver full color printing.

The great advantage of this technology is related to the coloration of the object. In this case, pigmentation occurs simultaneously with deposition, where a single nozzle head can result in different printing colors.

Previously, the manufacture of multi-color objects through the FDM process was only possible through multiple nozzles with multiple filaments, making the equipment complex and difficult to handle.

Figure 90 - Simplified representation of FDM process for manufacturing colored objects

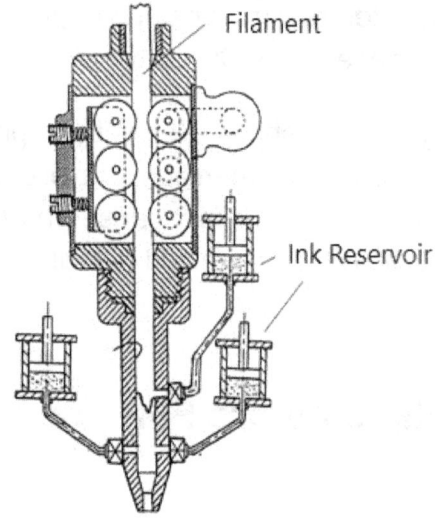

Source: From Patent (PRIDOEHL, SCHMITT ET AL. 2012)

A schematic of this new concept is shown in Figure 89, where ink reservoirs inject ink pigments into the plastic deposition nozzle via pistons.

In this way, the combination of colors injected into the deposition nozzle provides the deposition and printing of objects with color gradients.

Through this technique, color change over deposition is possible, although the transition between a new color and the previous color is not instantaneous.

An example of using this type of technology can be seen in Figure 91, where a color-transition object photo is presented. This project was firstly owned by BotObject, which was lately acquired by 3D Systems.

Figure 91 - FDM process photo for manufacturing colored objects

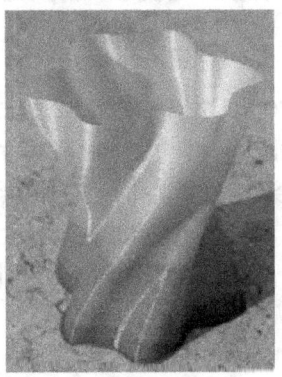

Source: (BOTOBJECTS 2014)

In this figure, you can see the use of 5 colors and transition gradients between these colors.

Through this type of equipment, it becomes possible to represent and manufacture models with higher mechanical strength than particulate 3DP equipment.

Similarly, pigments that modify the properties of the base material may also be used so that greater strength for only a more fragile part of the part is obtained.

5.3 Tendencies

Finally, we can see a global trend related to increased product differentiation and complexity, such as phones with higher processing capacity, stronger and lighter materials, parts with organic shapes based on nature shapes, among others. .

Similarly, there is a great effort to popularize additive processes (3D printers) to induce people to generate and materialize their ideas more often. Thus, by popularizing this kind of philosophy, it is possible to see the increase of entrepreneurship, since the paradigm of materializing an idea would be overturned or minimized.

We can also identify that customer-centric philosophy is intensified and product development process in mechanical and electronic engineering tends to adopt design thinking and agile methodologies. It is possible because each interaction will result in the deploy of a physical product made by 3D printing technologies.

As a consequence, we can also consider, in a long term, the increase of unsustainable widespread enterprises where the level of competitiveness would be high due to the high manufacturing cost and low production volume resulting from 3D printers.

For specific segments, we can foresee the strengthening of small and medium scale production based only on additive processes (3D printers), or mixed processes. In this way, high value-added, low-volume product ventures could become more competitive due to flexible 3D printer-based manufacturing systems.

It should be noted that in this case there is a maximum production volume where this type of system is profitable. Even so, as the product has a high manufacturing value, only high value added solutions and high level of innovation will be able to be based on the market.

Thus, for the economic survival of these new ventures, new markets need to be identified, besides new customer needs and value must be generated for customers.

In health care, a significant change is going to be in the way of working. The use of 3D printing technologies will be part of everyday life and will provide, for example, the construction of on-demand prosthesis using computed tomography in less than 1 hour.

Differentiated equipment solutions will also be developed, where CT equipment will already feature 3D printers to physically build exam results, assist in the analysis and generation of medical diagnostics more quickly and effectively.

With this, the time of diagnosis, surgery and recovery will be reduced for patients, providing increased life quality.

It can also be pointed out the possibility of manufacturing organs through 3D printers, so that incurable or difficult to treat diseases can be remedied.

It can also be predicted that houses and buildings will be built continuously, without the direct need for workers in ordinary activities. In this case, the main function of construction workers would be assigned to the noblest activities, such as finishing. Additionally, the use of 3D printing equipment for construction also minimizes the involvement of humans in hazardous activities.

Another important point in this matter is the use of these technologies in remote environments, such as space and seabed. Thus, the shape of buildings should occasionally increase their complexity.

Although there are many discussions and conjectures about such hypothetical situations, there are still several challenges to be overcome in 3D printers in order to consolidate these processes in scientific-industrial daily life.

References

3DSYSTEMS. Accura® 25 Plastic. 3D Systems. 2008a. Disponível em: <http://production3dprinters.com/sites/production3dprinters.com/files/downloads/DS_Accura_25_US.pdf>. Acesso em: setembro de 2011.

3DSYSTEMS. Accura® 48HTR Plastic. 3D Systems. 2008b. Disponível em: <http://production3dprinters.com/sites/production3dprinters.com/files/downloads/DS_Accura_48HTR_US.pdf>. Acesso em: setembro de 2011.

3DSYSTEMS. Accura® 55 Plastic. 3D Systems. 2008c. Disponível em: <http://production3dprinters.com/sites/production3dprinters.com/files/downloads/DS_Accura_55_US.pdf>. Acesso em: setembro de 2011.

3DSYSTEMS. Accura® 60 Plastic. 3D Systems. 2008d. Disponível em: <http://production3dprinters.com/sites/production3dprinters.com/files/downloads/DS_Accura_60_US.pdf>. Acesso em: setembro de 2011.

3DSYSTEMS. Accura® Amethyst Plastic. 3D Systems. 2008e. Disponível em: <http://production3dprinters.com/sites/production3dprinters.c

om/files/downloads/DS_Accura_Amethyst_US.pdf>. Acesso em: setembro de 2011.

3DSYSTEMS. Accura® Bluestone™ Plastic. 3D Systems. 2008f. Disponível em: <http://production3dprinters.com/sites/production3dprinters.com/files/downloads/DS_Accura_Bluestone_US.pdf>. Acesso em: setembro de 2011.

3DSYSTEMS. Accura® e-Stone™ Material. 3D Systems. 2009. Disponível em: <http://production3dprinters.com/sites/production3dprinters.com/files/downloads/DS_Accura_estone_US.pdf>. Acesso em: setembro de 2011.

3DSYSTEMS. Accura® CeraMAX™ Composite. 3D Systems. 2010a. Disponível em: <http://production3dprinters.com/sites/production3dprinters.com/files/downloads/DS_Accura_CeraMAX_US.pdf>. Acesso em: setembro de 2011.

3DSYSTEMS. Accura® PEAK™ Plastic. 3D Systems. 2010b. Disponível em: <http://production3dprinters.com/sites/production3dprinters.com/files/downloads/DS_Accura_PEAK_US.pdf>. Acesso em: setembro de 2011.

3DSYSTEMS. iPro™ 8000 & 9000. 3D Systems. 2011a. Disponível em: <http://production3dprinters.com/sites/production3dprinters.com/files/downloads/iPro-Family-USEN.pdf>. Acesso em: setembro de 2011.

3DSYSTEMS. ProJet™ 6000. 3D Systems. 2011b. Disponível em: <http://printin3d.com/sites/printin3d.com/files/downloads/Projet_6000_brochure_USEN.pdf>. Acesso em: setembro de 2011.

3DSYSTEMS. V-Flash. 3D Systems. 2011c. Disponível em: <http://printin3d.com/sites/printin3d.com/files/downloads/V-Flash_Brochure_USEN.pdf>. Acesso em: setembro de 2011.

ADDITIVE3D. 2. What are the limitations? Finishes. Castle Island Co. 2012. Disponível em: <http://www.additive3d.com/>. Acesso em: 01/01/2012.

BATCHELDER, J. S. SYRINGE TIP ASSEMBLY AND LAYERED DEPOSITION SYSTEMS UTILIZING THE SAME. PCT: Stratasys 2008.Disponível em: <http://v3.espacenet.com/textdoc?DB=EPODOC&IDX=WO2008130489&F=0>. Acesso em:

BENEATHTHEFALLEN. Pistol Frame Metalwork. 2010. Disponível em: <http://www.youtube.com/watch?v=nUtNPcyD-uU/

>. Acesso em: 15/01/2014.

BILLIET, T., M. VANDENHAUTE, et al. A review of trends and limitations in hydrogel-rapid prototyping for tissue engineering. Biomaterials, v.33, n.26, p.6020-6041.

BITSFROMBYTES. BFB 3000 Plus Brochure. 3D System. 2011. Disponível em:

<http://www.bitsfrombytes.com/>. Acesso em: agosto de 2011.

BOSSCHER, P. M.andL. W. I. I. ROBERT. Apparatus and method associated with cable robot system. USA: Ohio University 2007.Disponível em: <https://www.google.com/patents/US7753642>. Acesso em:

BOTOBJECTS. ProDesk3D Specification. BotObjects. 2014. Disponível em: <http://botobjects.com/specification/>. Acesso em: 01/05/2014.

CARRABINE, L. 3D Printers Deliver Competitive Alternative to Traditional Prosthetic Limbs. Make Parts Fast. 2010. Disponível em: <www.makepartsfast.com/>. Acesso em: 01/10/2012.

COOPER, K. G. Rapid Prototyping Technology Selection and Application. NY: CRC Press 2001.Disponível em: <http://dx.doi.org/10.1201/9780203910795.fmatt>. Acesso em:

COUTOURCRAFTING. Coutour Crafting Projetcs. USC - University of South Carolina. 2014. Disponível em: <http://www.contourcrafting.org/>. Acesso em: 01/10/2014.

CRUMP, S. S. Apparatus and method for creating three-dimensional objects. US: Stratasys, Inc. : 15 p. 1989.Disponível em. Acesso em:

CUNICO, M. M. W. M. Development of New Rapid Prototyping Process. Rapid Prototyping Journal, v.17, n.2, p.6-6. 2011.

CUNICO, M. W. M. Development of novel technology of additive manufacturing based on selective composite formation. (Thesis). Programa de Pós-Graduação em Engenharia Mecânica do Campus de São Carlos, University of São Paulo, São Carlos, 2013a. 329 p.

CUNICO, M. W. M. MÉTODO E APARATO PARA GERAÇÃO DE OBJETOS POR FUSÃO E DEPOSIÇÃO COM MULTIPLOS EIXOS. Inpi. BRASIL: Cunico, M. W. M. 2013b.Disponível em. Acesso em:

CUNICO, M. W. M.and D. A. KAI. Analysis of hybrid manufacturing systems based on additive manufacturing technology. Solid FreeForm Fabrication Symposium. Austin, 2017. p.

DEFDIST. Working Gun made with 3d Printing. Defense Distribuited. 2013. Disponível em: <http://defdist.org/>. Acesso em: 15/01/2014.

DINI, E. Method for automatically producing a conglomerate structure and apparatus therefor. USA: Enrico Dini 2008.Disponível em: <https://www.google.com/patents/US8337736>. Acesso em:

DOMINGOS, M., D. DINUCCI, et al. Polycaprolactone scaffolds fabricated via bioextrusion for tissue engineering applications. International journal of biomaterials, v.2009. 2009.

www.ingramcontent.com/pod-product-compliance
Lightning Source LLC
Chambersburg PA
CBHW080543220526
45466CB00010B/3017